THE BASICS OF CELL BIOLOGY

CORE CONCEPTS

THE BASICS OF CELL BIOLOGY

ANNE WANJIE, EDITOR

This edition published in 2014 by:

The Rosen Publishing Group, Inc.
29 East 21st Street
New York, NY 10010

Library of Congress Cataloging-in-Publication Data

Wanjie, Anne.
The basics of cell biology/Anne Wanjie.—1st ed.—New York: Rosen, © 2014
p. cm.—(Core concepts)
Includes bibliographical references and index.
ISBN 978-1-4777-0548-3 (library binding)
1. Cytology—Juvenile literature. I. Title.
QH582.5 W36 2014
571.6

Manufactured in the United States of America

CPSIA Compliance Information: Batch #S13YA: For further information, contact Rosen Publishing, New York, New York, at 1-800-237-9932.

CONTENTS

CHAPTER ONE

AN INTRODUCTION TO THE CELL

Cells are the building blocks of life. Your body contains trillions of cells, but many creatures are just a single cell.

Many areas of biology depend on an understanding of cells. The largest single cell, an ostrich egg yolk, is the size of a baseball, but cells can be so tiny that a hundred placed in a row would fit on the period at the end of this sentence. Tiny cells such as protists and bacteria are usually measured in nanometers. A nanometer is 1 millionth of a millimeter. Most plants and animals are made up of billions or even trillions of cells. The study of cells is called cytology.

Cells work together to carry out important functions within an organism. Many cells are specialized to carry out certain tasks. Red blood cells, for example, are specialized for transporting oxygen around the body.

Organisms like bacteria consist of just single cells. Each cell carries out all the processes it needs to survive.

Ostrich eggs contain the largest single cells on Earth. Each cell is the yolk cell that provides food for the growing ostrich chick. However, the yolk cells inside the eggs of many dinosaurs and some giant extinct birds such as moas were probably much larger.

THE CELL LIFE CYCLE

The cells that make up your body go through a life cycle. It consists of a period of growth followed by division to produce a new pair of cells. During the period of growth cells produce chemicals, make energy from food, and provide structural support for the organism. Communication between the different parts of a cell and between a cell and its neighbors is very important to ensure they work together for the benefit of the organism.

Cells contain structures such as the cell membrane, cell wall (in plants and bacteria only), and a series of fibers called the cytoskeleton. They support the cell. Movement often depends on hairlike extensions of the cell. Long single extensions are called flagella. Groups of shorter projections that beat rhythmically are called cilia.

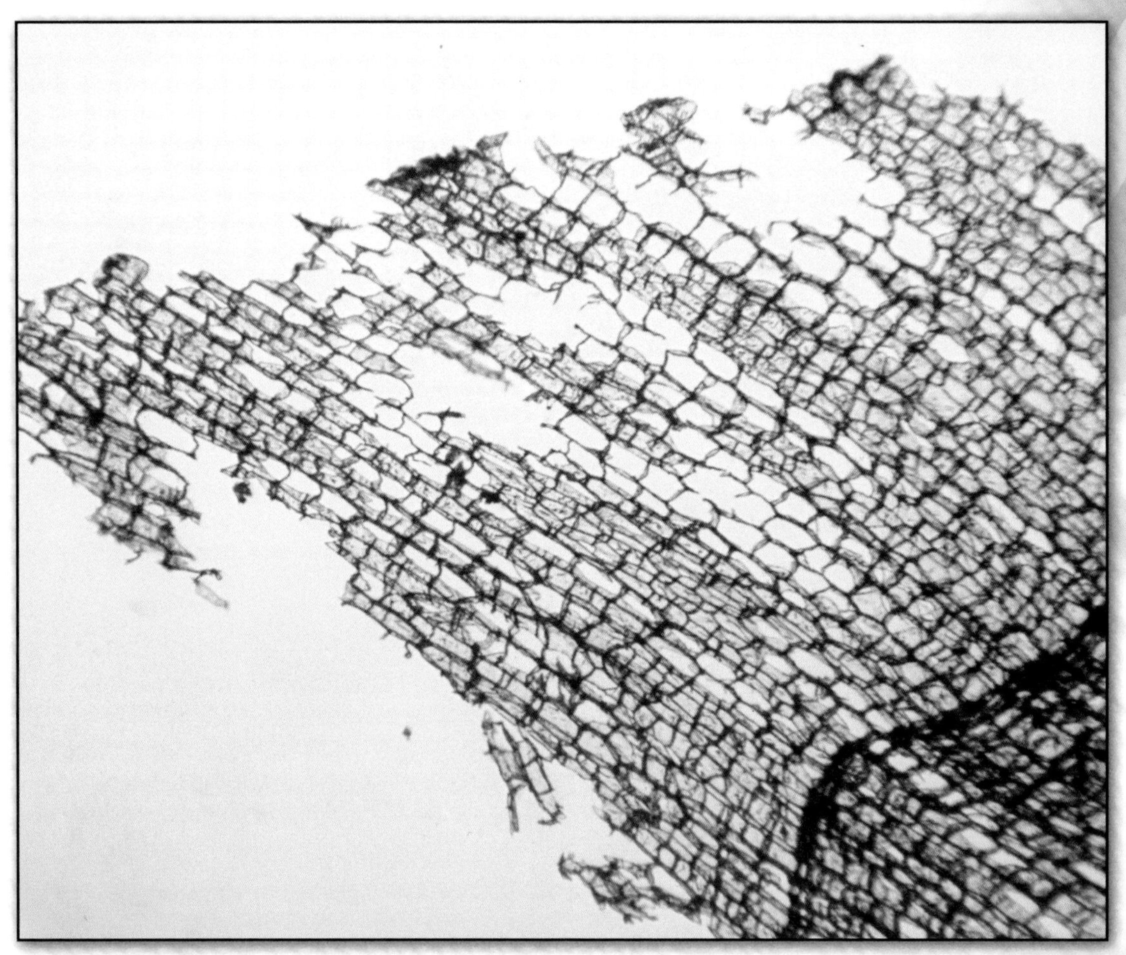

A slice of cork under a light microscope. Robert Hooke would have seen something similar as he peered through his microscope in 1667.

VIEWING CELLS UNDER A MICROSCOPE

The first microscope was made by a Dutch eyeglass-maker, Zacharias Janssen (1580–c.1638). Later scientists improved on this early version. Englishman Robert Hooke (1635–1703) used his microscope to look closely at a thin slice of cork. Hooke used the word *cell* to describe the units he could see, since they reminded him of the small rooms called *cells* that monks lived in.

Hooke believed the cork cells were empty and that the walls were made of living material. Improvements in microscope design enabled Dutch scientist Anton van Leeuwenhoek (1632–1723) to study the cellular world in far greater detail. Leeuwenhoek was the first person to see protists, blood cells, and sperm. He also noted the presence of some cell contents such as the chloroplasts, though he was unable to describe them accurately.

In the nineteenth century microscope technology advanced further. Cytologists (cell biologists) could fully observe cells and their contents. Study of the very finest detail became possible with the invention of the electron microscope in the 1950s.

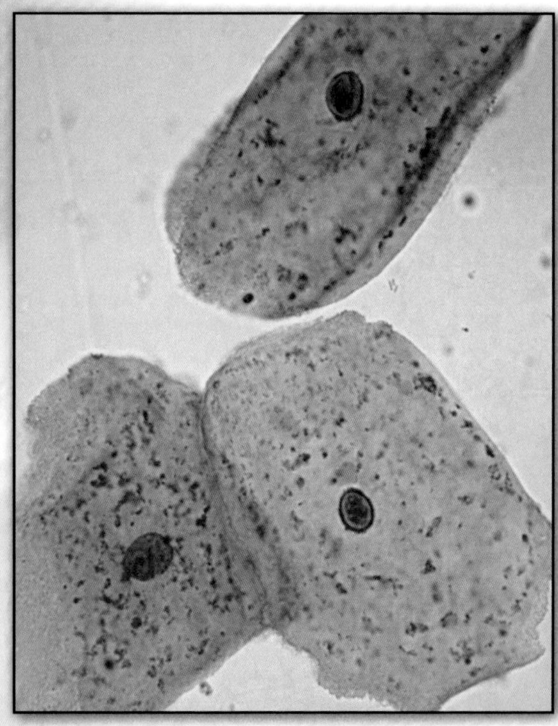

These are cheek cells. They are easy to collect from the inside of a person's cheek. Cheek cells are often the source of DNA used for DNA fingerprinting, an important tool in police work.

WORKING TOGETHER

Similar cells work together and become specialized in their functions to form tissues, such as muscles or blood in animals. A number of tissues working together make up an organ, such as a heart, stomach, or kidney. Groups of organs work together to form systems, such as the digestive system.

A NEW THEORY

Although cells were discovered and named by Robert Hooke, the idea that cells form the basic unit of living organisms was established by German biologists Matthias Schleiden (1804–1881) and Theodor Schwann (1810–1882). This became known as the cell theory. Schleiden studied plant cells, and Schwann studied animal cells. Later, German physician Rudolf Virchow (1821–1902) showed that all cells arise from preexisting cells through cell division.

PROKARYOTES AND EUKARYOTES

Cytologists divide cells into two types, prokaryotes and eukaryotes. Prokaryotes are creatures such as bacteria. They are single-celled organisms, although some occur in chains or clusters of many thousands of individuals.

The DNA of a prokaryote cell floats freely in a region called the nucleoid. The rest of the cell is called the cytoplasm. It contains a thick, jellylike liquid called cytosol and tiny structures called ribosomes. Ribosomes use instructions encoded in the DNA to produce proteins. The cell is wrapped by a cell (plasma) membrane and, in many cases, a tough wall, too.

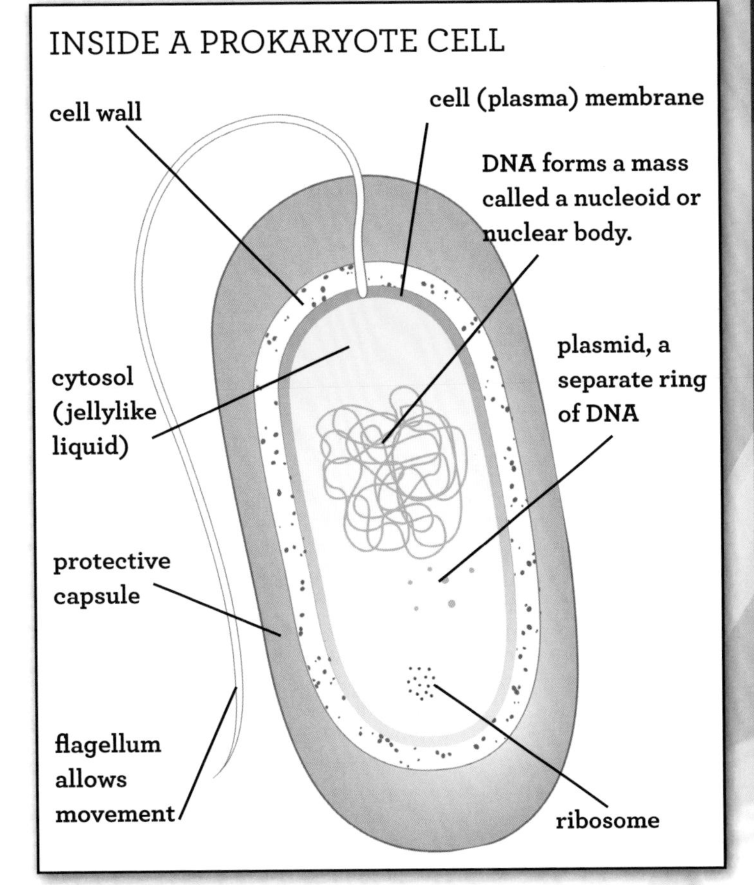

This is a bacterial cell. There is great diversity of bacterial form and function, but all have ribosomes, a cell membrane and wall, and DNA loose in a nucleoid, which is not enclosed by a membrane.

ENDOSYMBIOSIS

Scientists believe that the mitochondria that occur inside eukaryote cells descend from free-living prokaryotes. This theory is called endosymbiosis. Evidence comes from the fact that mitochondria have their own DNA separate from the nucleus.

Around 1.5 billion years ago a prokaryote engulfed the mitochondrion ancestor. Rather than being digested, the mitochondrion stayed alive, providing energy to the other cell and getting a safe place to live in return. Over millions of years of evolution the two cells became inseparable.

Eukaryotes probably evolved from such a union of cells. Chloroplasts are also thought to be the result of an ancient endosymbiosis.

INSIDE PLANT CELLS

Take a celery stalk, and put one end in some water with some blue ink or food coloring added. After an hour remove the celery and rinse it. Then cut the stalk into pieces, and examine the cut ends. You will see tiny spots of color. They are sections through bundles of cells that run through the stalk called xylem. They carry water from the roots to the leaves.

How far has the colored water traveled in one hour?

EUKARYOTES

Like prokaryotes, eukaryote cells have a cytoplasm, cell membrane, and ribosomes. But they are usually much larger, and they contain many other features absent in prokaryotes. They include a series of fibers called the cytoskeleton. It moves materials around and maintains cell shape. There are also membrane-bound structures called organelles. They do much of the cell's work. The nucleus is the largest organelle. It contains DNA, which is the cell's genetic (inherited) information.

Other organelles include the endoplasmic reticulum, which packages proteins; lysosomes, inside which large molecules are broken down; the mitochondria, which produce energy from food; and, in plants, algae, and some bacteria, chloroplasts, which harvest the sun's energy to make food.

SIZE MATTERS

As an object gets bigger, its volume increases more quickly than its surface area does. Look how the ratio of surface area to volume drops with size in these cubes. The same thing happens with cells. They need a high surface area-to-volume ratio to work efficiently. That is why most cells are tiny. It also explains why large organisms must be made up of many small cells rather than just a few giant ones.

PLANT AND ANIMAL CELLS

Plant and algal cells have strong walls that lie over their cell membranes. Plant cells also contain a large sac called a vacuole, which can make up more than 70 percent of the volume of the cell. The vacuole takes in water and begins to swell. That pushes the cytoplasm against the cell wall, making the cell rigid and giving the plant stability and strength.

The cells of fungi such as mushrooms are similar in some ways to plant cells. They usually contain vacuoles, and their cells have tough walls. However, fungal cells do not have chloroplasts. That is because, like animals, fungi do not photosynthesize. Instead, fungi get the energy they need by breaking down dead and decaying material.

Animal cells are usually smaller than plant cells. The main difference between the structures of animal and plant cells is that animal cells do not have vacuoles or tough cell walls. The membranes of animal cells are made of a flexible material, so the cells can change their shape and size easily. For movement and support animal cells produce bones, cartilage, or shells and tissues like muscles.

PROTISTS

The most complex single cells belong to the eukaryote group called protists. Protists are a group of incredible diversity. Some photosynthesize, while others hunt. Protists move in many different ways, but many stay in one place. Protist features include light receptors, sensory bristles, poison darts, leglike body extensions, and even bundles that contract just like muscles.

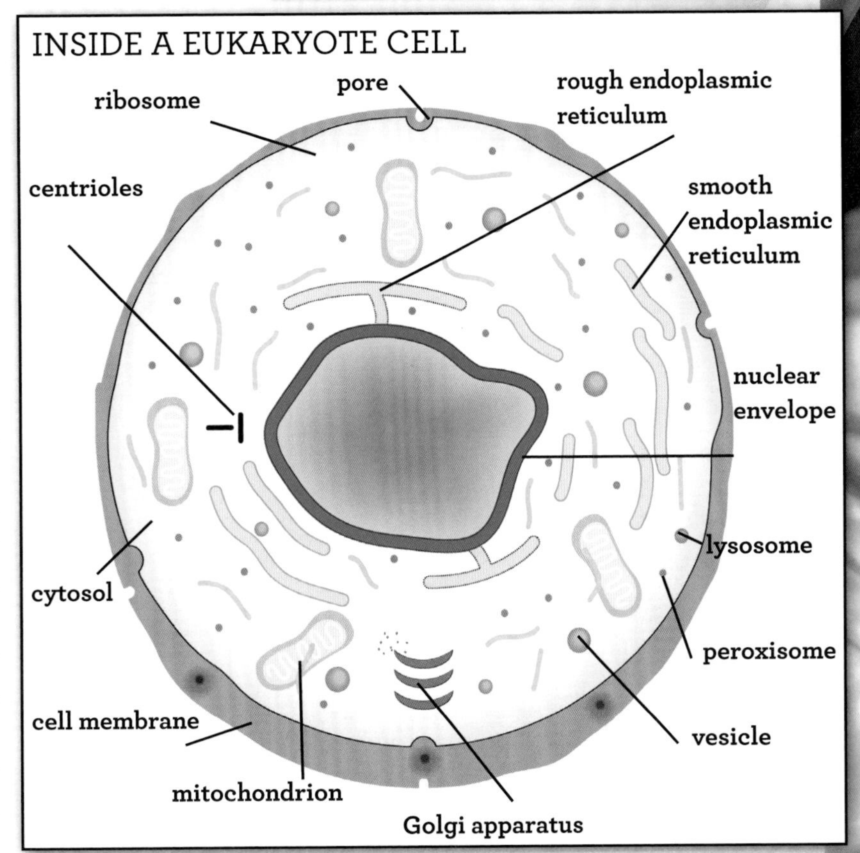

This is an animal cell. Plant cells are a little different. They have a tough cell wall and organelles called chloroplasts that trap the energy of sunlight.

CHAPTER TWO

A VARIETY OF TYPES

There is tremendous diversity among the cells that form tissues and organs in organisms.

A plant or animal can be made up of trillions of cells. Different cells carry out different tasks. A tough chemical, cellulose, occurs in plant cell walls, giving them rigidity. By contrast, animal cells are flexible, so they can change size and shape. Groups of cells that work together to carry out certain functions form tissues. Some tissues, such as the pith of plants, are made up of cells of the same type. Other tissues are complex mixtures of many different types of cells. For example, many types of plant cells are required to form structures such as flowers, leaves, and seeds.

Animals contain a range of different types of cells too. Multicellular (many-celled) animals contain structures that provide support, such as shells, exoskeletons, or bones. Cells forming these structures are called support cells. Organisms also contain lining and nerve cells, and often muscle cells, too.

These disk-shaped cells are red blood cells. Formed in the bone marrow, red blood cells shuttle oxygen and carbon dioxide around the body. Blood is a type of connective tissue.

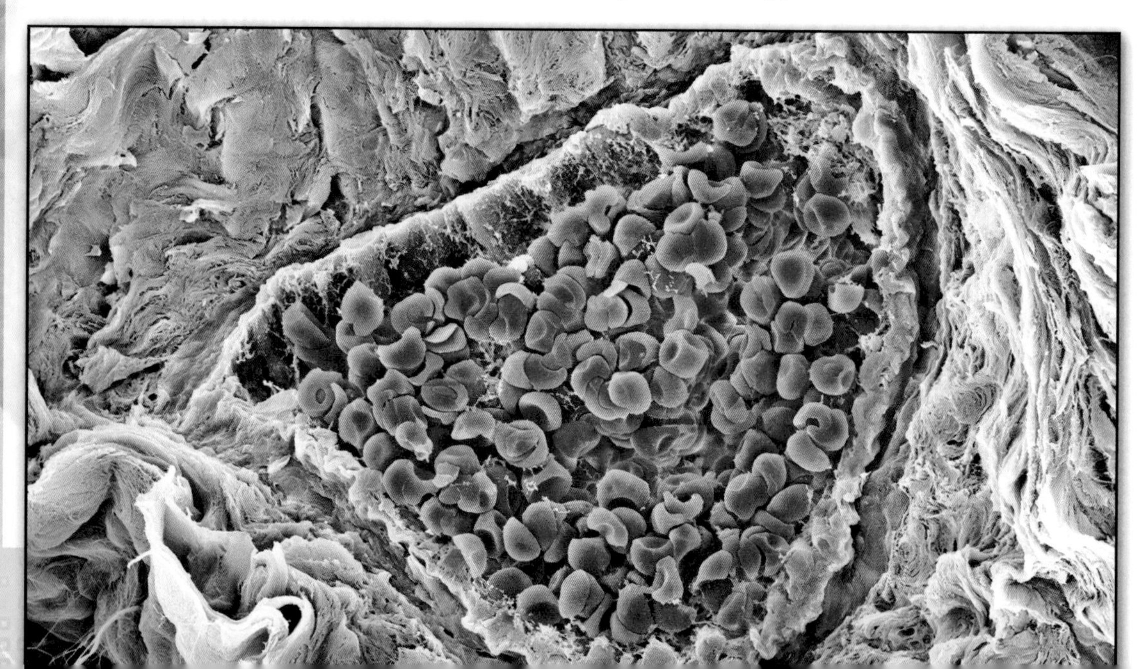

INSIDE PLANT CELLS

Turgor pressure allows these pitcher plants to stand tall. However, if the rate of water loss through the leaves exceeds the rate of water uptake at the roots, the plants lose turgor pressure, then wilt.

Plant cells are often larger than animal cells. Plant cells contain large water-filled packages called vacuoles. The water inside exerts a force on the membrane enclosing the vacuole. This is called turgor pressure. It causes cells to become turgid (rigid). Turgid cells tightly packed together produce a very firm type of tissue. It provides the main means of support for plants other than those strengthened by wood. When nonwoody plants are deprived of water for a time, the plant loses turgor pressure. The vacuoles shrink, and the structural tissues lose their firmness. Soon the plant begins to wilt.

Structural tissues called parenchyma form the bulk of nonwoody plants. Parenchyma cells in the pith, deep inside the plant, are not exposed to light. They are of no use for photosynthesis, the process of making sugars using carbon dioxide gas and sunlight. So pith parenchyma cells lack chloroplasts,

CELLS FOR REPRODUCTION

The sex cells—sperm and eggs—are specialized single cells. Sperm are tiny. They have a whiplike tail called a flagellum that they use to get around. They have mitochondria for energy, but very little cytoplasm. All a sperm contributes to a zygote (fertilized egg) is its genetic material. Egg cells are much larger. They contain food supplies for the zygote as well as a supply of all the organelles it will need as it develops.

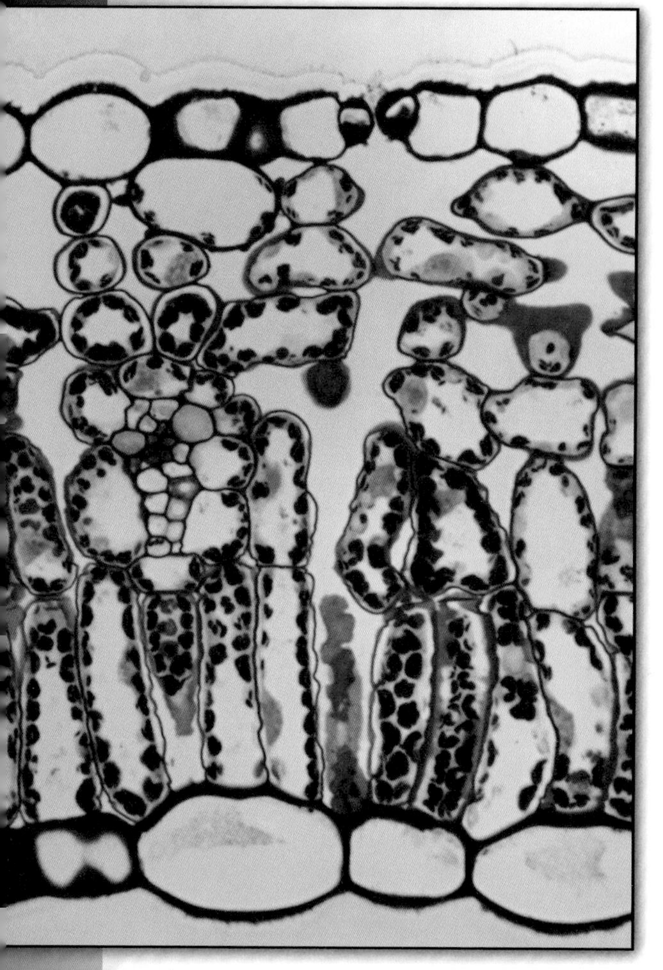

This is a section through a lilac leaf. The horizontal cells at the top are epithelial cells. The vertical cells are parenchyma; they contain chloroplasts.

THE GENOME

All the cells in an organism result from many divisions of a single fertilized egg cell. With the exception of sex cells, all the cells in an organism are genetically identical. How, then, can they perform different functions?

Different cells use different parts of their genome (genetic instructions). For example, a pancreas cell uses genes that code for digestive juices. Although genes that code for other features, such as bone repair or hair growth, are present, they are not expressed by these cells. The characteristics of a certain cell depend on stretches of DNA called regulator genes. They function as switches and can turn other genes on or off.

the miniorgans inside which photosynthesis occurs.

THE CELLS IN LEAVES

Chloroplasts do occur in the parenchyma cells of leaves. The outermost layer of the leaf is called the epidermis. Epidermal cells produce a waxy coating on their

THE STRUCTURE OF LEAVES

Take a look at the leaf surfaces of different plants in your backyard. Many plants show specializations in their epidermis. For example, they might have hairs or spines, or have a thick, waxy outer layer. How do you account for the different shapes and colors of leaves from different plant species? Thicker leaves have broad parenchyma layers and may also have thick waxy layers. These features might prevent plants from losing too much water in dry environments. Darker leaves may have more chloroplasts. The plant might grow in a shady place, so it needs more chloroplasts to make the most of the reduced light.

outer cell walls. It acts as a barrier against water loss, as well as disease organisms and insect enemies. Epidermal cell walls must be thin, though; otherwise not enough light can reach the chloroplasts inside.

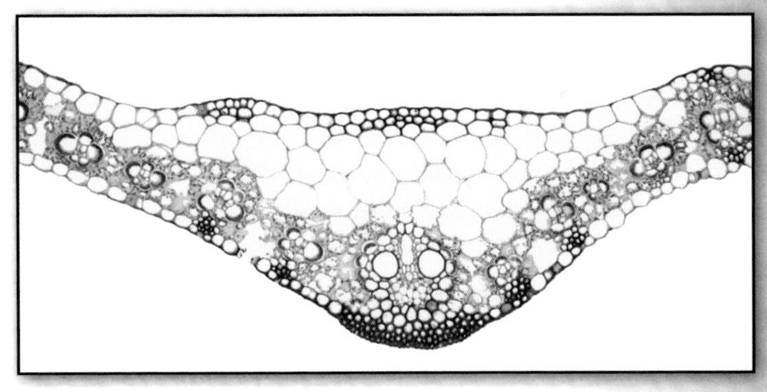

A section through a leaf rib. Xylem cells are in the center, surrounded by a thin layer of phloem. It is skirted by parenchyma cells, with epithelial cells around the outside.

PROVIDING SUPPORT

Other plant cells have thicker walls. They provide support for the plant. Support cells occur wherever extra strength is needed, in the leaf stalk, for example. Collenchyma cells are stretchy support cells. Their walls become so thick that the cell itself dies. These dead cells form the tough outer layers of the shells of nuts and seeds.

TRANSPORTING WATER

The thick-walled cells that strengthen a plant also form long transport vessels. Xylem tissue is formed of hollow, dead cells. Water is drawn into the plant's roots. It then moves up the xylem to the leaves. There water is lost through a process called transpiration that is caused by water evaporating from pores in the leaves into the air. Transpiration sucks the water up through the xylem. Xylem also forms wood in trees. Sugars and other molecules also need to be carried through the plant. Dissolved in water, they

CROSS-SECTION OF A PLANT STEM

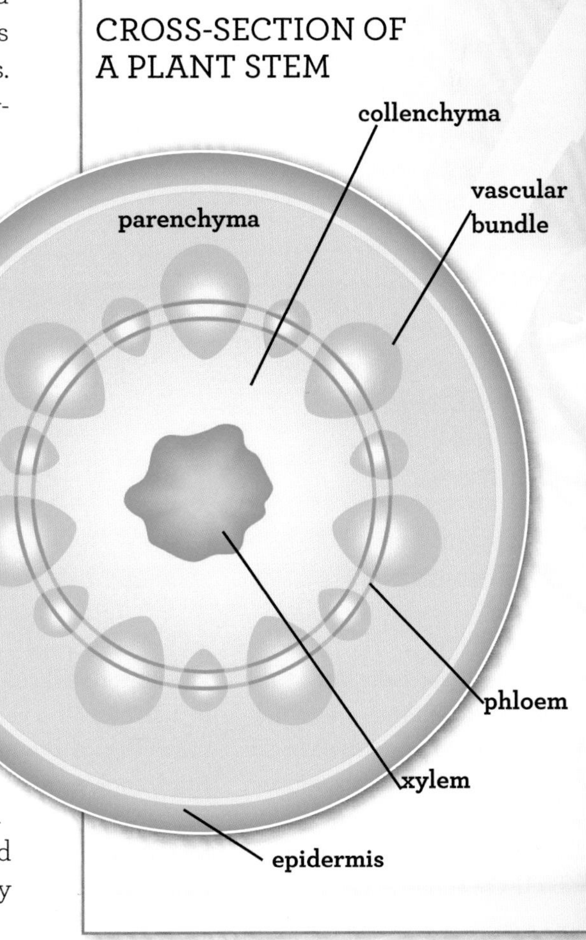

move around the plant through a different system of cells called the phloem. Unlike dead xylem cells, phloem cells are alive. A mesh of tiny pores allows the contents of one phloem cell to flow into the next.

ANIMAL TISSUE

Scientists divide animal tissues into four main groups depending on the way they develop in the embryo. They are lining, supporting, muscle, and nerve cells.

EPITHELIAL CELLS

Epithelial (lining) cells join to form sheets that cover the surface of the body, as well as the outer surfaces of organs. The inner surfaces of organs are lined by endothelial cells.

Cells on the skin's surface divide rapidly to replace cells that are sloughed (brushed off). Epithelial cells that line the air sacs of lungs form a tissue just one cell thick. It allows oxygen to pass easily into

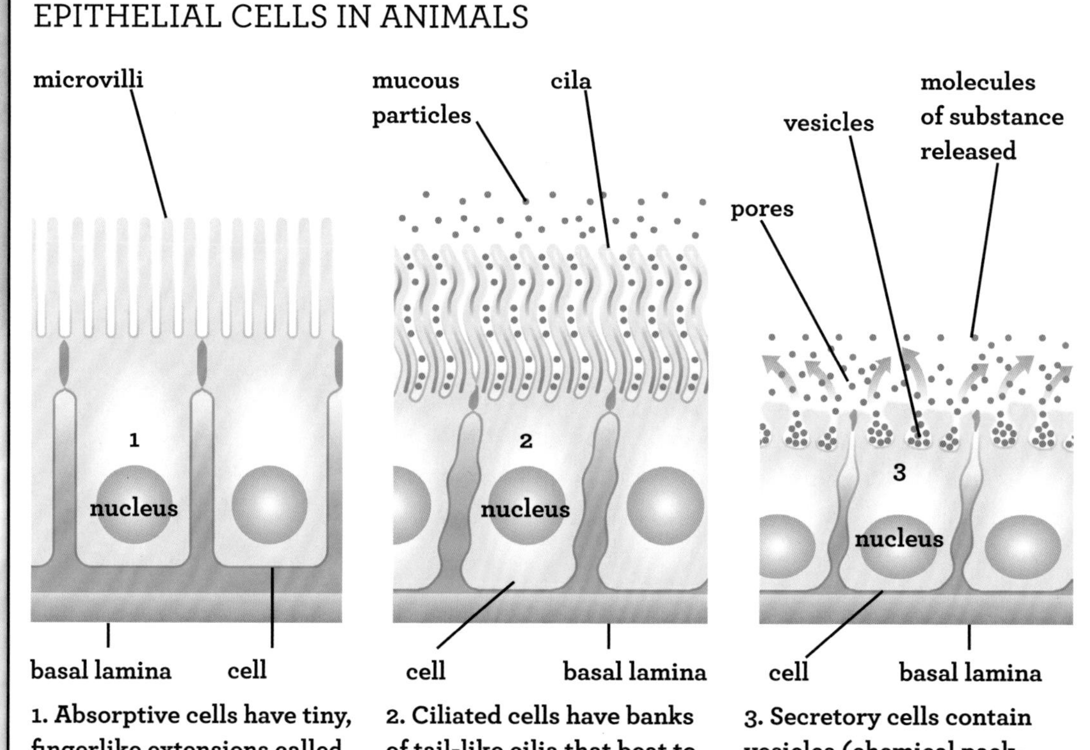

1. Absorptive cells have tiny, fingerlike extensions called microvilli. They increase the area available for absorption of molecules enormously. The walls of the gut are lined by these types of cells.

2. Ciliated cells have banks of tail-like cilia that beat to move things suspended in fluid, such as mucus. Mucus in the windpipe is moved by ciliated cells. Cilia help many tiny organisms get around.

3. Secretory cells contain vesicles (chemical packages). The vesicles rupture to release a substance onto a body surface, such as the skin or into the gut. Sweat glands, for example, are lined by secretory cells.

the blood and carbon dioxide to move out. Cells lining the windpipe are shaped like columns. They are covered with tiny projections called cilia. The cilia wave in unison to send mucus to the back of the throat, where it is swallowed.

Cells line small sacs called glands. Glands release products such as hormones. Glands in the gut release chemicals called enzymes that digest food. Sweat glands are tubes that empty onto the skin. During exercise the glands fill with water and dissolved salts. These constituents of sweat pass into and through epithelial cells lining the sweat glands.

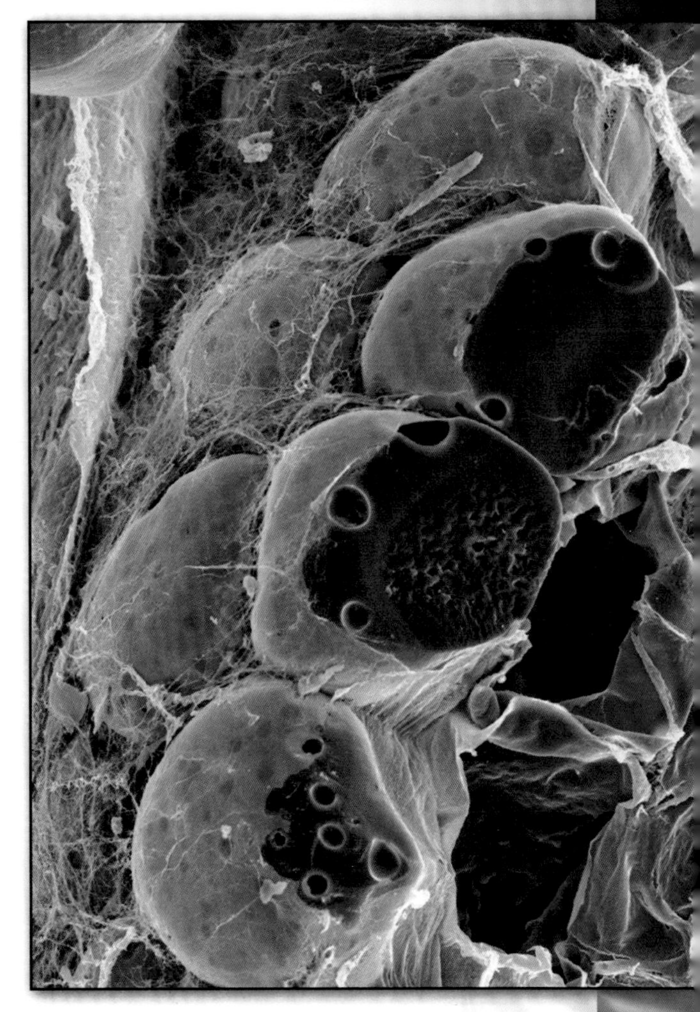

Adipose, or fat, tissue is a type of connective tissue. The liquid-filled cells serve as energy supplies, cushion the internal organs, and insulate the body against the cold.

CONNECTIVE TISSUES

Tissues important for packing and support are called connective tissues. Cartilage, bone, and blood are all types of connective tissues. Their cells are usually widely separated and are surrounded by complex mixtures of materials, such as collagen fibers. This kind of loose tissue occurs around all the organs of the body. It also connects the skin to structures directly beneath it. Denser connective tissue containing more fibers occurs where greater strength is needed, such as in ligaments and tendons. Harder

CELLULAR ERRORS

Sometimes errors occur in cell differentiation. If white blood cells start to multiply uncontrollably, a kind of cancer of the blood called leukemia develops. There are two types of white blood cells, granulocytes and lymphocytes. Similarly there are two types of leukemia. The type of disease depends on which blood cells are multiplying.

A DISCOVERY IN DINOSAUR BONES

In 1974 French anatomist Armand de Ricqles made an exciting discovery. De Ricqles looked at thin slices of fossilized dinosaur bone and found large numbers of Haversian canals inside. They are channels that form in bone around blood vessels. Haversian canals occur today in fast-growing animals like mammals and birds, but not in reptiles like lizards. This suggested that dinosaurs grew very quickly to adulthood.

More provocatively, the study suggested that dinosaurs were warm-blooded like mammals, not cold-blooded like lizards and other reptiles as most biologists believed at the time. Later research seems to have proven that dinosaurs were warm-blooded. Studies of oxygen in Tyrannosaurus rex bones proved that the temperatures of these dinosaurs varied little in life, as in mammals and birds. Further evidence came in the 1990s with the discovery in China of dinosaurs with insulating coats of feathers.

types of connective tissues include cartilage and bone. Bone gets its toughness from its structure—layers of cells sandwiched between layers of tough minerals. Materials like bone are called composites.

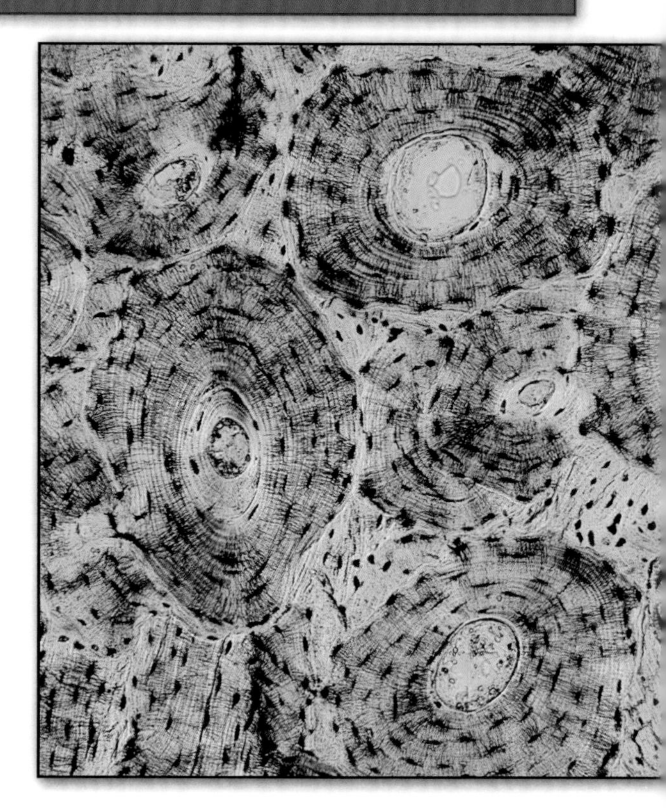

This is a section through mammal bone. The hole in the center is a Haversian canal. Blood vessels run through it. They keep the living tissue of the bone supplied with oxygen and nutrients, and take away waste.

MUSCLE TISSUE

The cells of muscle tissue are able to contract (shorten) and relax, allowing movement. Muscle cells can do this because they contain bundles of fibers that slide over one another.

Muscle contraction is a process that demands a lot of energy, so muscle cells are packed with mitochondria, the organelles that release energy from food.

NEURONS

Nervous tissue has many densely packed nerve cells called neurons. There are at least ten billion neurons in the human brain. Neurons carry electrical signals

INSIDE A NERVE CELL

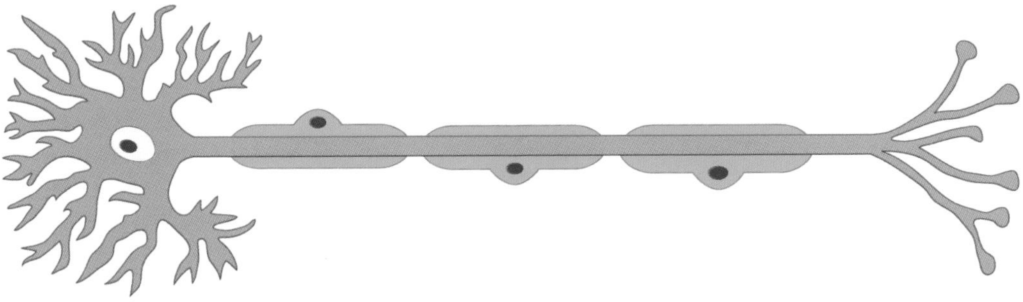

The structure of a motor neuron, a nerve cell that triggers activity in muscles, organs,and glands. Myelin insulates the neuron, while the nodes of Ranvier act as signal amplifiers.

between the brain and body. This electrical activity depends on the movement of tiny particles called ions. Each neuron consists of a central cell body that contains the nucleus (control center) and other organelles. There are several extensions of the neuron. Most cells have one long extension called an axon; they also have shorter ones called dendrites. The extensions can carry electrical messages over great distances. Nervous tissue also contains supporting cells, which can be ten times more numerous than the neurons. Supporting cells store food reserves, fight infection, or insulate the nerve cells with fatty coatings.

CELL DIVERSITY

Most animals have bodies composed of tissues, organs, and systems, but some have much simpler arrangements. The smallest animals, placozoans, are little more than bundles of cells. Sponges are larger, but they, too, do not have distinct types of tissues.

Sponges live in the ocean, attached firmly to the seabed. A system of water channels runs through the body of a sponge. Cells with flagella line the insides of the channels. They beat their flagella to create a current that draws in fresh water from outside. The sponge cells then filter tiny particles of food from the water. The water current then runs out through

NEURON INSULATION

The myelin sheath insulates neurons against electrical interference from other nerve cells. It also prevents signals from degrading as they travel along the neuron. The sheath is peppered with gaps called nodes of Ranvier. Each gap measures less than one thousandth of a millimeter across. Electrical impulses can jump across the gaps, allowing faster travel; in this way the nodes amplify the nerve signals.

a large opening at the top of the sponge called the osculum.

Cnidarians—a group of animals that includes corals, jellyfish, and sea anemones—have more diverse cell types. Unlike sponges, cnidarians have epithelial cells and muscle cells that help at least some of their life stages move around. However, cnidarians lack many of the cell types found in other animals, such as blood and a brain.

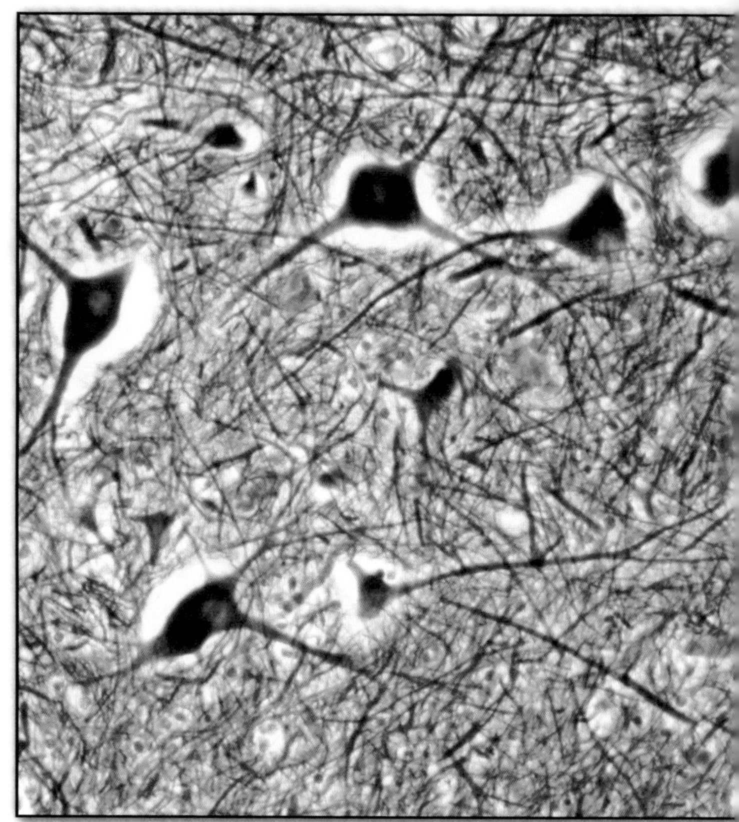

A section through brain tissue composed of nerve cells.

UNDERSTANDING STEM CELLS

Unspecialized cells in the body are called stem cells. A stem cell can become any other type of cell. Stem cells occur in embryos before the tissues and organs develop. The stem cells then begin to differentiate (specialize for certain functions). It is not just embryonic cells that differentiate. Adults also have stem cells, such as the cells that divide to become blood cells. Stem cells can divide almost without limit and can be grown outside the body. That provides medical researchers with a window of opportunity, although the research is far from complete. For example, stem cells could be forced to differentiate into pancreas cells. They could then be implanted into the pancreas of a diabetes sufferer. Dopamine-producing cells could treat Parkinson's disease, while Alzheimer's, heart disease, and cancers may also become treatable. However, adult stem cells are not ideal for medical use. Stem-cell research depends on taking cells from embryos. The embryos used are "spares" left over from IVF programs. Stem cell research remains highly controversial, since many people think the use of these cells is abhorrent.

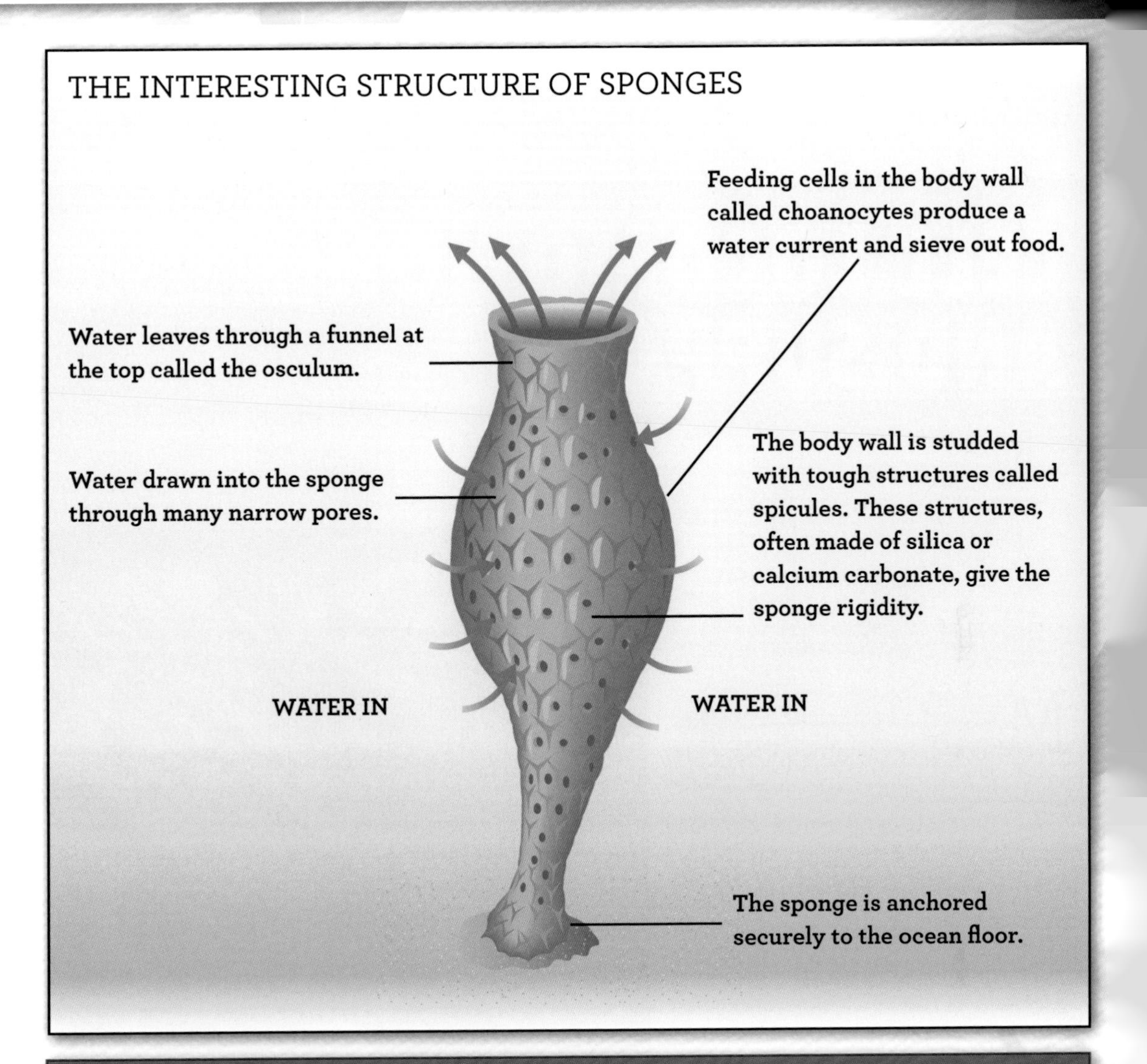

WHAT ARE CNIDARIANS?

Cnidarians catch and kill other animals for food. A unique type of cell called a cnidocyte helps them do this. Each cnidocyte contains a structure called a nematocyst. It is a capsule attached to a tightly coiled thread with a sharp barb at the tip. When the cnidocyte is touched by prey, a lid covering the cell pops open. The thread then explodes out of the cell, and the barb drives into the prey. Poison then goes through the barb to paralyze the prey, leaving it unable to move. The prey is then drawn into the gut for digestion. Sea slugs are mollusks, so they are unable to produce nematocysts. However, some sea slugs take nematocysts from their prey and use them for their own defense. Aeolid sea slugs are immune to the poisons of the cnidarians they eat. The slugs store the stolen nematocysts in a pouch and use them to fend off predators.

CHAPTER THREE

CELL LOCOMOTION AND SUPPORT

Cells have internal structures that support them, control their movements, and carry out functions.

Many cells have structures that allow them to move. There are also parts that move inside a cell. Other structures provide support and keep the cell's shape. An essential structure shared by all cells is the cell membrane. Surrounding the cell, it gives support as well as a barrier to the cell's environment.

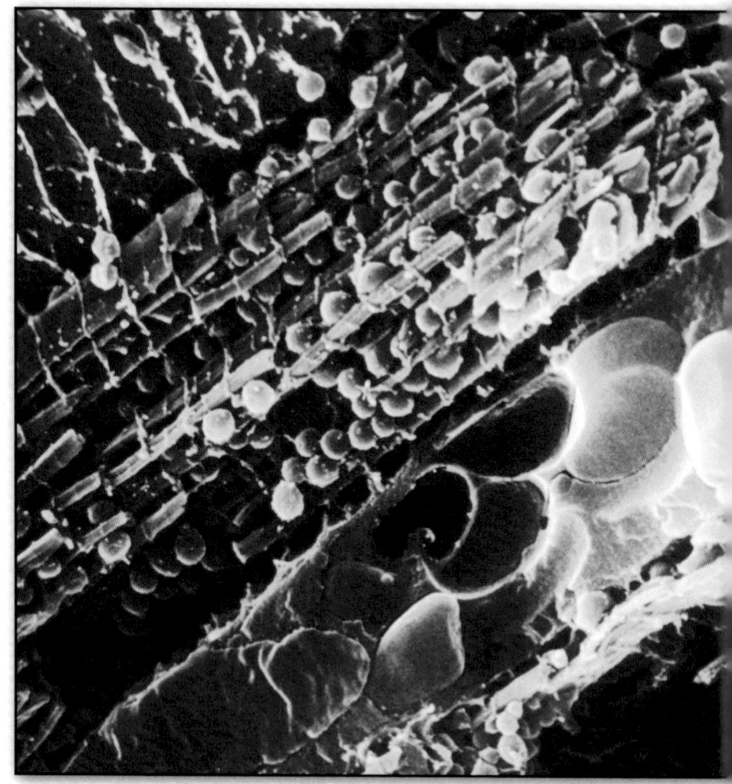

Human muscle filaments. Muscle contraction involves the movement of filaments containing the proteins actin and myosin.

THE CELL MEMBRANE

The cell membrane is very thin and flexible, but it is strong, like a layer of cling wrap. It separates the inside of the cell from the outside, stops the contents from spilling free, and keeps other chemicals out. Most of the membrane is composed of complex fat molecules called phospholipids.

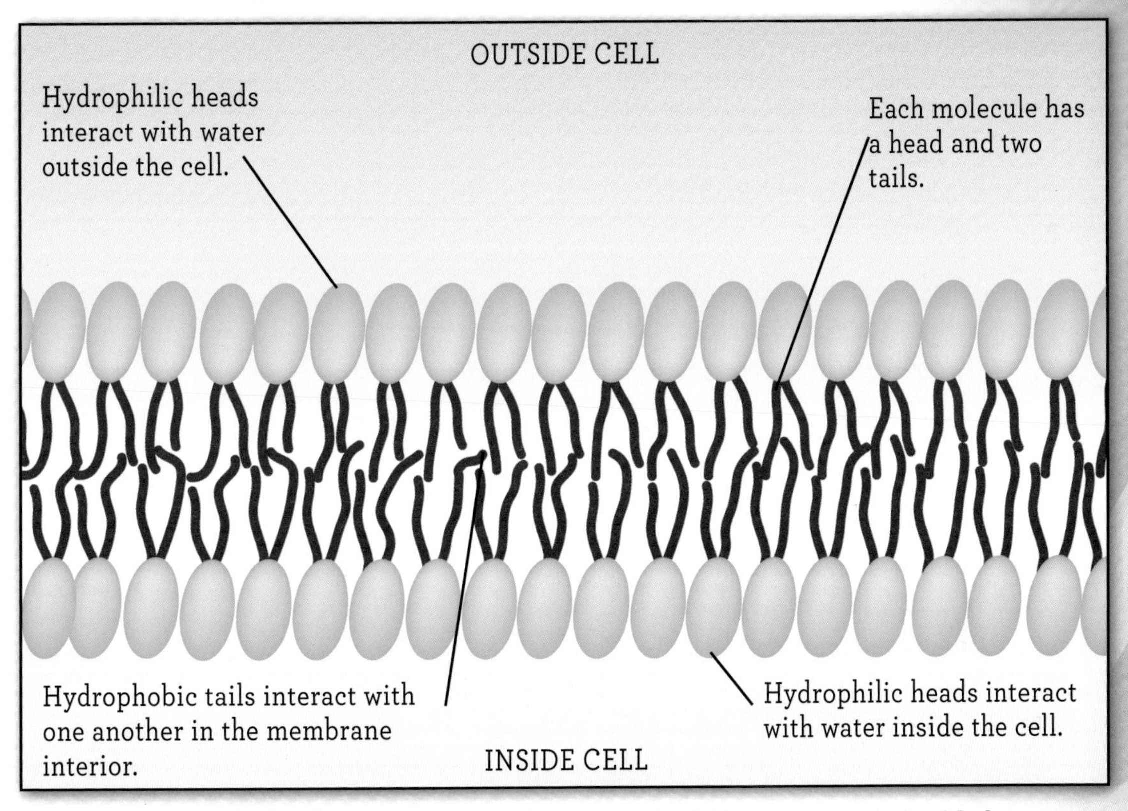

Different phospholipid sections respond to water in different ways. A double layer (or bilayer) can separate two water-containing areas, such as the inside and outside of a cell.

Cell membranes depend on the structure of phospholipids to function. Some chemicals are hydrophobic, meaning they repel water. Others attract water; they are called hydrophilic. Phospholipids are both. They have a pair of "tails" composed of fatty acids. The tails are hydrophobic, so they repel water. The "head" of a phospholipid molecule contains phosphorus; it is hydrophilic.

Phospholipids in a membrane are arranged in two layers. The tails link on the inside, while the two sets of heads pack together closely to face outward. The two layers are called the lipid bilayer.

The cell membrane gives structural support to the cell, but it is also flexible. It serves to limit the movement of molecules in and out of the cell. Important chemicals that the cell needs to draw in or force out move through large proteins embedded in the membrane. These proteins form a series of gates and channels that allow chemical transport. Other membrane proteins provide anchorage for an internal support structure called the cytoskeleton.

MEMBRANE PROTEINS

The cell membrane does not consist solely of phospholipids. It is also studded with molecules called membrane proteins. There are usually around 25 phospholipids for each protein. Membrane proteins extend across the lipid bilayer. Like phospholipids, they have water-attracting sections extending away from the membrane and water-repelling sections that link inside it. Some membrane proteins float around within the bilayer, although many are anchored by the cell cytoskeleton. The proteins have a range of functions, such as shuttling molecules in and out of the cell or acting as receptors for chemicals released by nerves.

The cell membrane also contains carbohydrates. They act as recognition sites for other cells, allowing the cells to stick together.

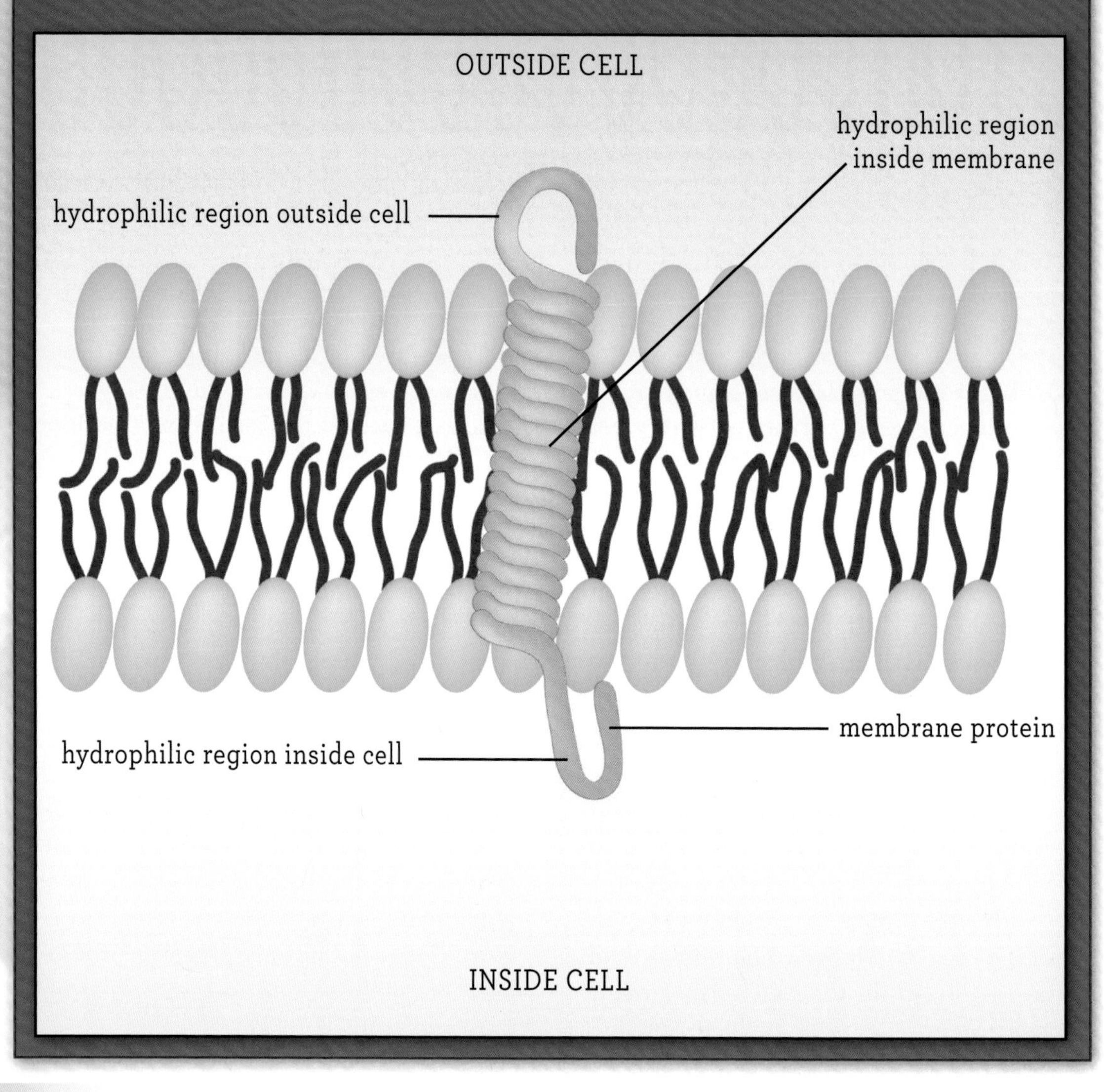

PLANT CELL WALLS

The cells of plants are more rigid than animal cells. That is because plant cells have a tough cell wall in addition to the cell membrane.

Plant cell walls consist of a type of sugar called cellulose. The cellulose is embedded in a network of other chemicals. The cell wall gets its strength from the way its cellulose molecules connect. Bundles of around 250 molecules align to form a structure called a microfibril. Other, smaller sugars form tough bridges between the microfibrils. In order to grow, cell walls must expand. A growth hormone (messenger chemical) called auxin causes the cell wall to soften. It then expands in a direction determined by the way the microfibrils line up.

The cell wall is rigid so it can resist pressure from water inside the cell. Without a cell wall the cell would burst. Instead, it becomes rigid, like a balloon

PLANT CELL EXPANSION

Plant cells are lined by tough banks of cellulose-rich microfibrils. How, then, do they grow? Plant cells grow when a hormone called auxin binds to receptors inside it. That makes the cell wall lose its toughness. It becomes stretchy, allowing the cell to start expanding. However, the direction of the cell's expansion depends on which way the microfibrils run around the cell wall.

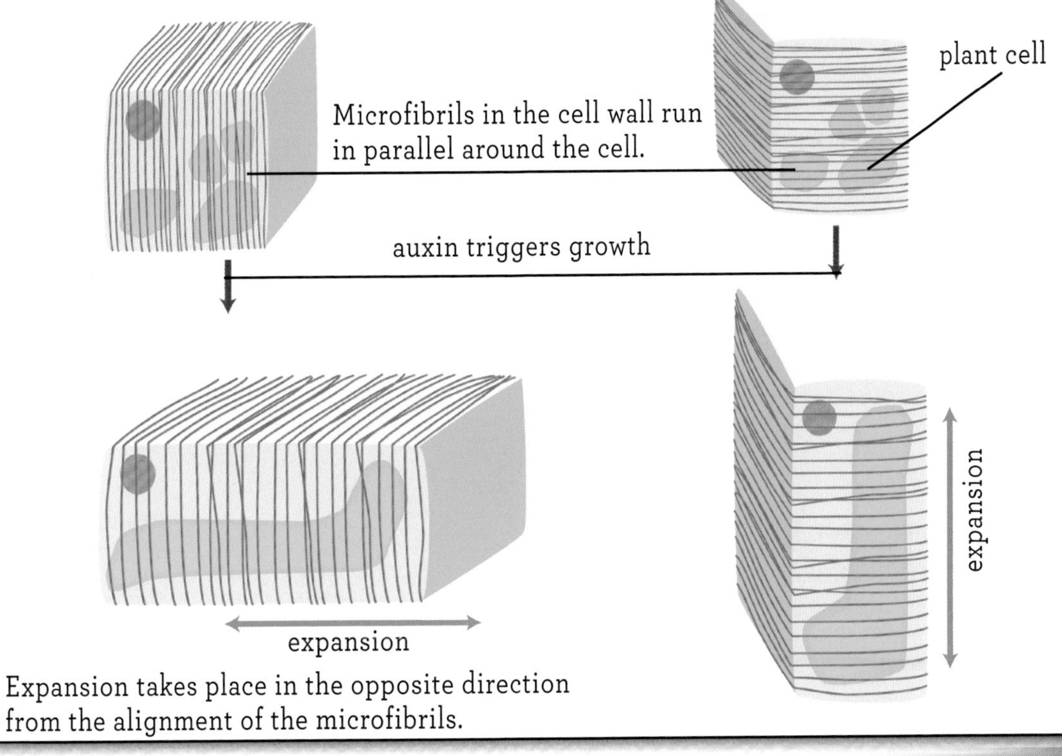

Expansion takes place in the opposite direction from the alignment of the microfibrils.

BACTERIA CELL WALLS

Like plants, bacteria have cell walls. Bacterial cell walls, though, are much more varied. Their main functions are to maintain the shape of the cell, to prevent its contents from drifting away, and to keep out unwanted chemicals. Scientists use a technique called Gram staining to detect the presence of certain bacteria, including ones that cause disease, like *Listeria* and *Streptococcus*. Bacteria that can be detected in this way are called Gram-positive bacteria. The cell walls of Gram-positive bacteria contain large amounts of substances called peptidoglycans. They react with certain dyes to produce a deep violet color. Antibiotics like penicillin work by disrupting bacterial cell walls. Once its cell wall is breached, a bacterium cannot retain the concentrated soup of chemicals its needs to function, and it dies.

filling with air. That allows well-watered plants to grow tall and strong even if they do not contain hard woody tissues.

CELL CONNECTIONS

Cell membranes and walls connect cells to their neighbors. There are several types of cell junctions. Chemical messages go from cell to cell through gap junctions. Tight junctions prevent molecules moving along the spaces between cells.

Desmosomes are junctions that provide structural support. They occur in

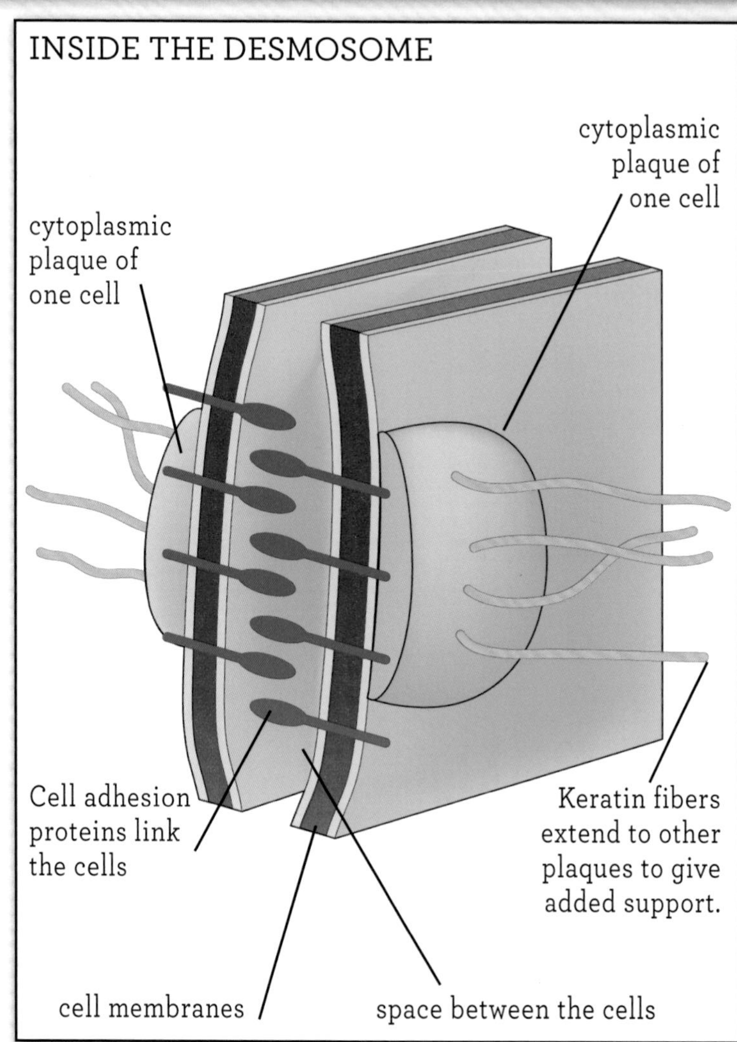

A desmosome, a tough type of joint that links two cells. Desmosomes occur in the skin and other places of regular wear and tear.

SPINDLE FORMATION

One of the best-known microtubule functions is the formation of the spindle during cell division. The spindle is a set of microtubules that appears when a cell is ready to divide. First, microtubules of the regular cytoskeleton break down. Units of tubulin are then reassembled into a structure shaped like a birdcage. That is the spindle. It grows longer and longer, forcing the cell to stretch. At the same time, molecules of DNA (which carry genetic information) coil to form chromatids, which pair up to form structures called chromosomes. Microtubules from the ends of the spindle attach to the chromosomes. The microtubules then shorten, pulling each chromatid from its partner, which is pulled in the opposite direction. Once the DNA is divided, the cell divides, and the spindle disintegrates. A new cytoskeleton then forms in each of the new cells.

epithelial tissues, which lie on the outside of body surfaces, forming the skin or wall of the gut, for example. Desmosomes are very tough joints. Proteins called keratin extend through the cell cytoplasm from desmosome to desmosome. Keratin also forms structures such as the fingernails.

CELL SUPPORT

Prokaryotes depend on the cell wall and membrane for support. Eukaryote cells such as those of animals and plants contain a network of fibers that provides support. It is called the cytoskeleton. It

MICROTUBULES

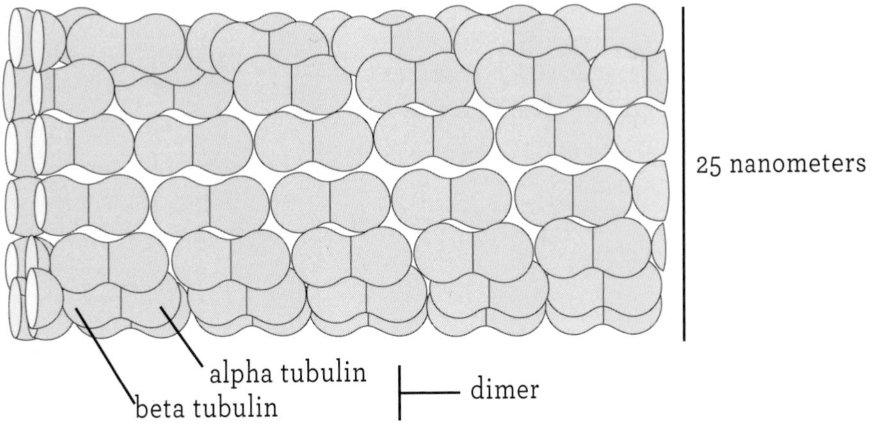

Microtubules are hollow cylinders formed by the protein tubulin. There are two types of tubulins, alpha and beta, that link together to form a molecule called a dimer. Microtubules grow or shorten by the addition or removal of dimers

NANOMETERS

A nanometer is a tiny measurement indeed. It is equal to one billionth of a meter, or around one 25-millionth of an inch.

criss-crosses the cytoplasm and underpins the cell membrane. There are three main types of fibers in the cytoskeleton. They are microtubules, actin filaments, and intermediate filaments.

ABOUT MICROTUBULES

The largest fibers in a cell's cytoskeleton are the microtubules. At 25 nanometers across, these fibers are just large enough to be seen with an ordinary light microscope.

Microtubules have a diverse range of functions. They are hollow tubes made of a protein called tubulin. Every microtubule contains 13 tiny filaments. They weave together to make the complete fiber. The fiber can be lengthened or shortened by adding or taking away tubulin molecules.

ACTIN

At just 7 nanometers across, microfilaments are the smallest of the fibers that occur in the cytoskeleton. They are made of a protein called actin. A microfilament consists of two actin polymers that are twisted together to form a helix, or spiral, shape. Just like microtubules, microfilaments can be rapidly lengthened or shortened by adding or taking away actin molecules. That is important for changing the cell's shape and for generating movement.

Actin is one of the most abundant proteins in an animal's body. Almost 10 percent of your soft tissues (excluding water) is actin.

INTERMEDIATE FILAMENTS

Intermediate filaments are somewhere between microtubules and microfilaments in diameter. They are different from other cytoskeletal fibers because their exact composition varies with the type of cell. Intermediate filaments are also more stable and form a permanent scaffold inside the cell—their sole

TYPES OF INTERMEDIATE FILAMENTS

Intermediate filaments from different tissues vary in their chemical makeup. Six types of intermediate filament proteins occur in animals. Usually only one occurs in any one tissue type. Scientists can find out the tissue in which a cell originated by studying the intermediate filaments inside. That is called intermediate filament, or IF, typing. IF typing is of great importance in the fight against cancer. Cancer tumors can quickly spread around the body. By looking at the intermediate filaments of a cancerous cell, scientists can decide where in the body the cancer started. That helps doctors target treatments more effectively.

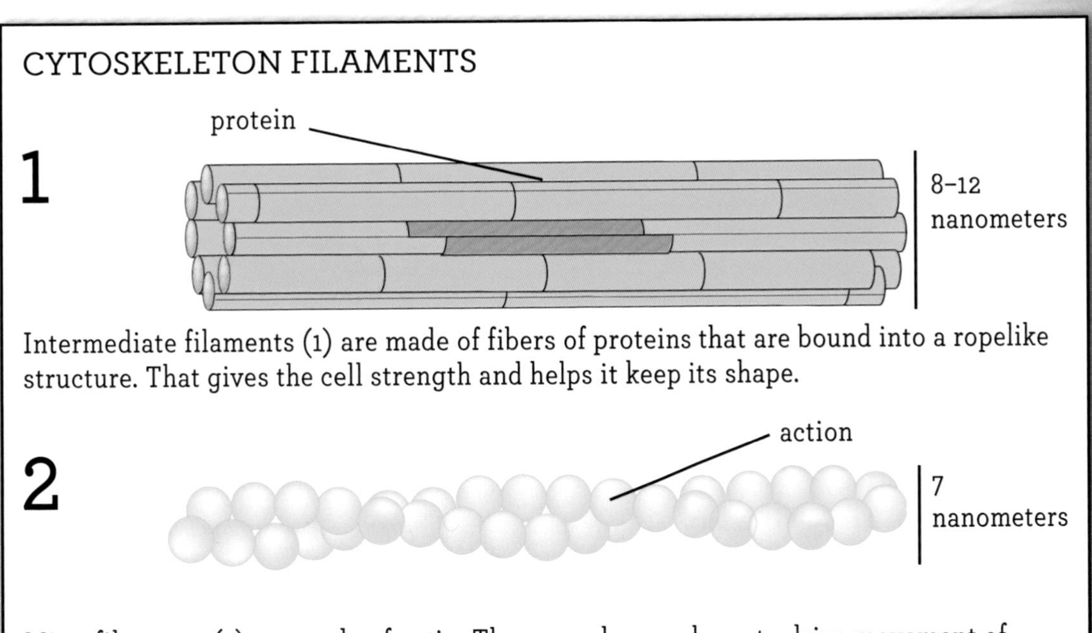

Intermediate filaments (1) are made of fibers of proteins that are bound into a ropelike structure. That gives the cell strength and helps it keep its shape.

Microfilaments (2) are made of actin. They can change shape to drive movement of other cellular structures. Microfilaments and myosin work together to allow muscle contraction.

function is to provide internal support. Intermediate filaments give cells their strength.

AMEBA MOVEMENT

As well as providing structural support, the cytoskeletal proteins play an important role in the movement of cells as well as whole organisms. Other types of proteins, such as myosin and dynein, are also important for movement.

HOW AMEBAS MOVE AROUND

Amebas are protists that move around in an incredible way. The cytoplasm of an ameba contains a liquid, endoplasm, that is usually near the center of the cell. The cytoplasm also contains a more solid substance, ectoplasm, that lies close to the cell membrane. For the cell to move, the ectoplasm pushes out to form a bump. The endoplasm then begins to flow into the bump. This forms an extension called a pseudopod, or "false foot." As the liquid

FLAGELLA

You can see cells moving with an ordinary light microscope, the kind in your school biology lab. Take a sample of ordinary pond water, and see if you can spot some single-celled organisms. How do they move? See if you can figure out whether the creatures are using cilia, flagella, or ameboid locomotion to swim around.

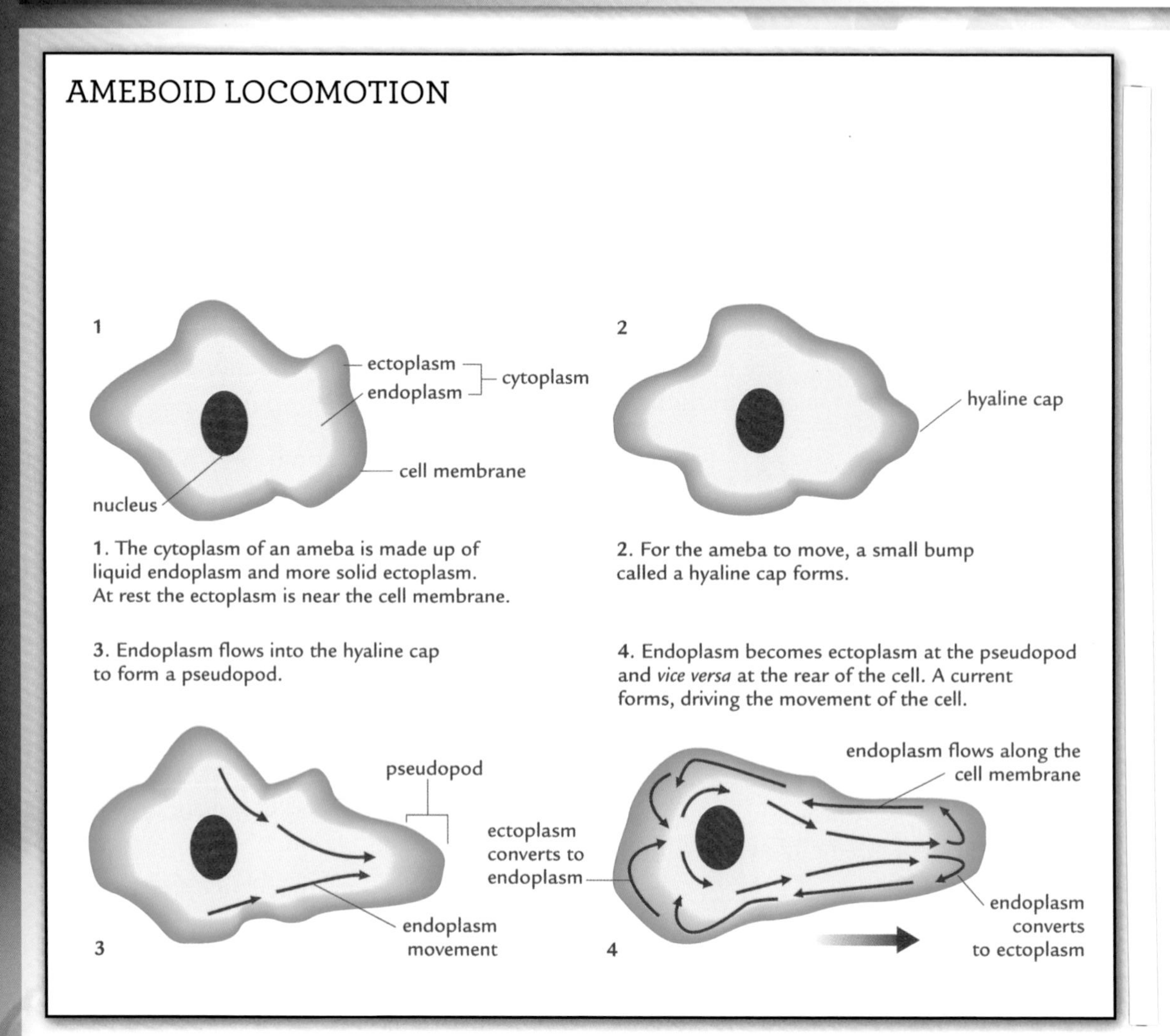

1. The cytoplasm of an ameba is made up of liquid endoplasm and more solid ectoplasm. At rest the ectoplasm is near the cell membrane.

2. For the ameba to move, a small bump called a hyaline cap forms.

3. Endoplasm flows into the hyaline cap to form a pseudopod.

4. Endoplasm becomes ectoplasm at the pseudopod and *vice versa* at the rear of the cell. A current forms, driving the movement of the cell.

endoplasm reaches the false foot, it is converted into more solid ectoplasm and runs back along the cell membrane.

At the same time, the opposite takes place at the rear of the ameba. There ectoplasm changes to endoplasm, which flows through the cell to the false foot. In this way the ameba slowly inches forward. Biologists call this type of movement ameboid locomotion.

LOCOMOTION

Many protists use ameboid locomotion to get around. Others use a more active system. These creatures move by using cilia (sing. *cilium*) or flagella (sing. *flagellum*). Cilia and flagella are tail-like extensions of cells. Flagella are long and move in a whiplike fashion. Usually there are just one or two on a cell. A wave runs down the flagellum, driving the cell along. Cilia

CILIA MOVEMENT

1. The cilium beats against the liquid around it. This is called the power stroke. The resistance of the liquid (or drag) creates a force. The sum of forces from many cilia push the creature forward relative to the liquid.

2. To return to its start position, the cilium adopts a different stroke. It is called the recovery stroke. It minimizes the drag so a force is not produced that pushes the creature back the way it came.

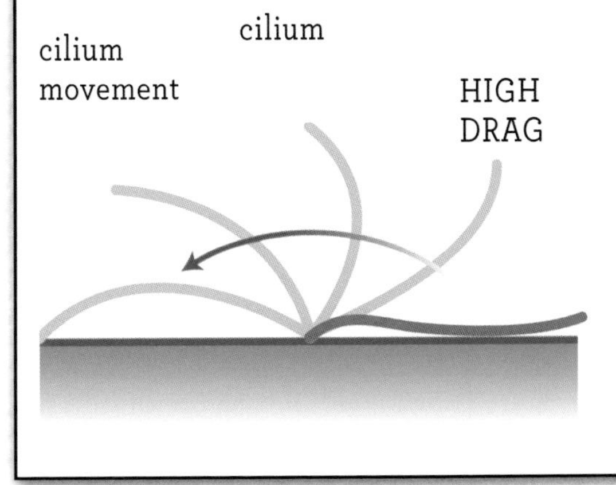

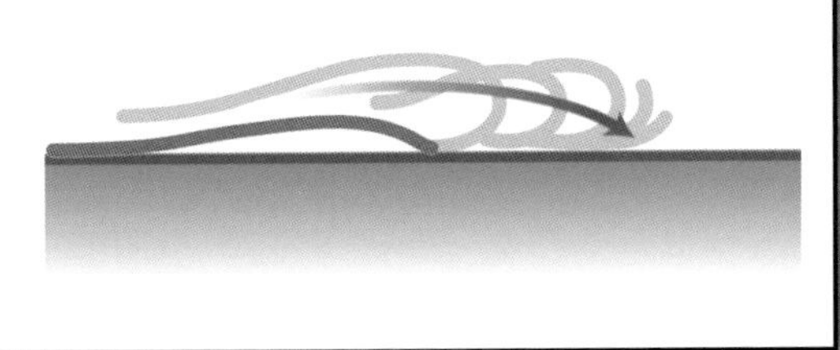

Cilia can move a tiny creature through water or help move a liquid, such as mucus, in a larger animal.

are shorter and usually occur in much greater numbers. Individual cilia are less effective than a flagellum; but when they beat in concert with many others, they make good oars. Cilia move in a way similar to a swimmer doing the breaststroke; a power stroke is

FLAGELLA MOVEMENT

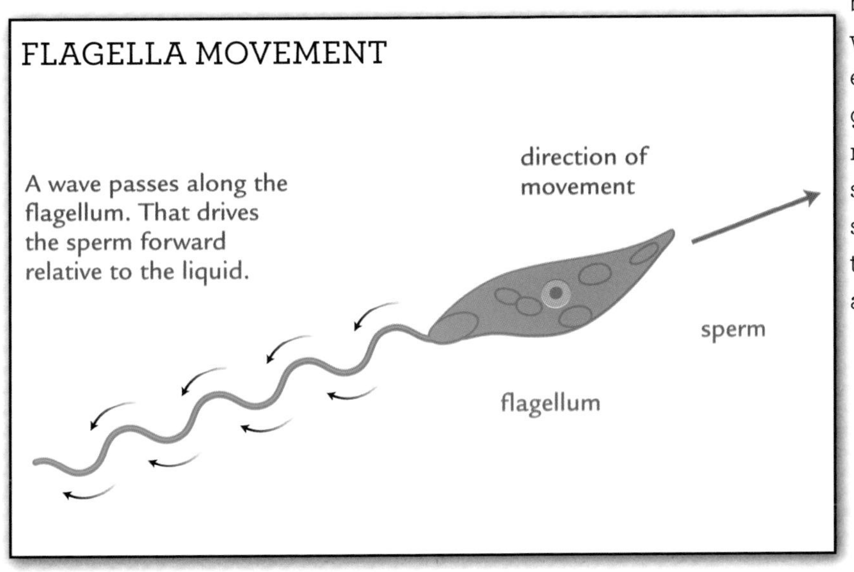

How a flagellum, such as one on a sperm, pushes a cell forward through a liquid.

BACTERIA

Many bacteria also have flagella. However, they are very different from those of eukaryotes such as protists, suggesting that their flagella evolved independently. Bacterial flagella do not have microtubule doublets or even dynein. They are made of a different protein, flagellin. Bacterial flagella do not have waves of motion; instead, they spin around like an airplane propeller

followed by a recovery stroke, in which the cilium keeps as close to the cell membrane as possible.

THE MECHANICS OF CILIA AND FLAGELLA

Cilia and flagella anchor onto the cytoskeleton. Although they look different, their internal structures and the way they power their movements are the same. A cilium consists of nine pairs (or doublets) of tiny tubes called microtubules. One of each pair is complete; the other is incomplete and is fused to its partner. The doublets are arranged in a circle to form a larger tubular structure.

The doublets surround a single central pair of tubules, which are bound by a sheath. Biologists call this the "9 + 2" arrangement. Nine protein spokes radiate from the sheath to the outer pairs. Movement depends on the doublets sliding past each other. To do this a motor protein called dynein is needed. Dynein forms "arms" on each doublet that attach to the

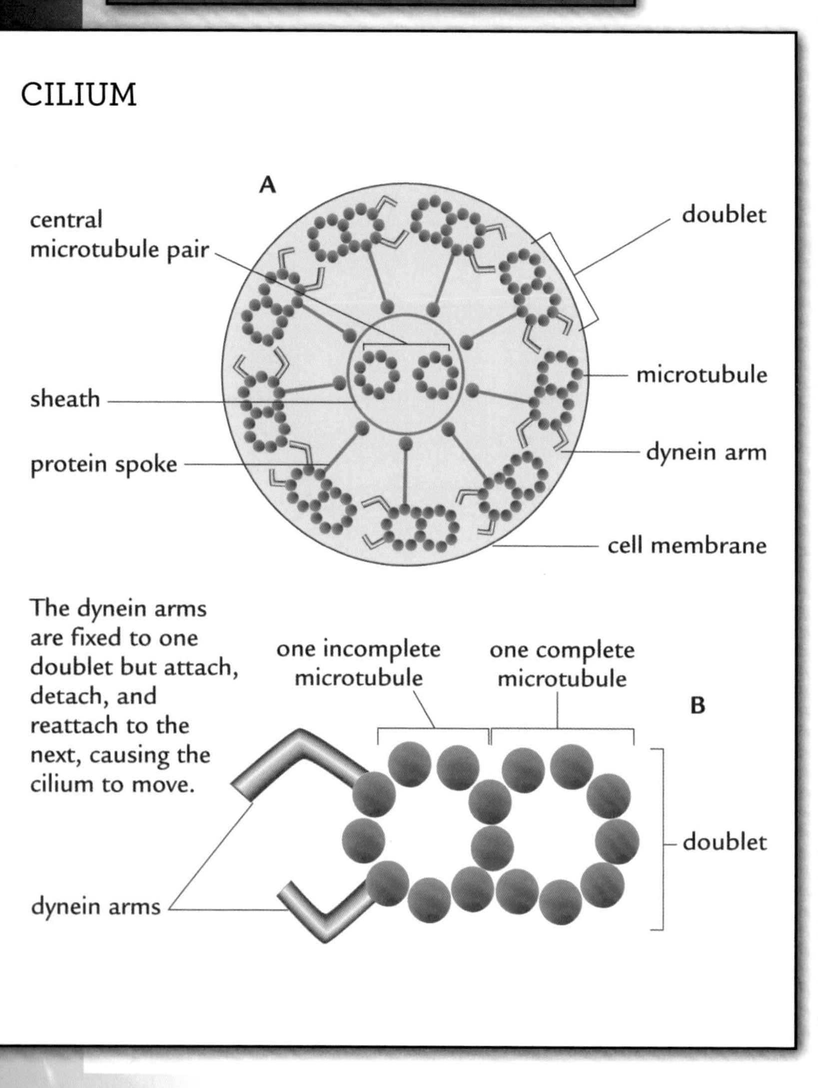

A cross-section through a cilium (A) shows the 9 + 2 arrangement of microtubules. B shows a doublet and the motor protein, dynein, in greater detail.

next doublet along. A chemical called ATP changes the shape of a dynein arm. That makes the next doublet move.

Many tiny creatures use cilia for propulsion. The largest ciliated animals, comb jellies, are around 0.5 inches (1.25cm) across. Movement using cilia is inefficient for larger animals, but cilia are still important for moving things around inside the body. For example, mucus in the windpipe is moved along by beating banks of cilia. Similarly, animal sperm swims by using flagella. For the whole animal to move, though, muscle is needed.

MUSCLES MOVEMENT

Muscle function is similar to ciliary movement, since filaments attach and reattach to slide across each other. Muscle is formed of bundles of fibers called myofibrils. Each contains sections called sarcomeres. They are made of actin microfilaments that overlap with filaments of another protein, myosin. Myosin filaments are thick, and each is surrounded by six thinner actin microfilaments.

To contract the muscle, nervous signals cause the release of calcium ions. That triggers the myosin molecules to attach to actin. The attachment causes the filaments to move a little (around 10 nanometers). An ATP molecule then binds to the myosin, making it release the actin. The myosin is now ready to reattach to the actin. By repeating this cycle many times, the sarcomere shortens, and the muscle contracts.

RIGOR MORTIS

When an organism dies, its muscles soon stiffen. This is called rigor mortis. Death stops the transport of ATP into muscles. The bonds between actin and myosin cannot be broken, so the muscles stiffen. Eventually the proteins begin to break down, and the muscle softens. These events take place on a predictable timescale. Police scientists study the stiffness of a corpse in a homicide enquiry. That can help the scientists establish the time that has passed since the victim's death.

The structure of a myofibril in detail. Sarcomeres are the units that contract to make muscles move. Light filaments in the sarcomere contain actin; darker filaments are composed of myosin.

CHAPTER FOUR

A LOOK INSIDE THE CELL

Cells can be described as factories. Each has the machinery needed to perform its own particular tasks and keep itself healthy.

The internal structure of cells varies with cell type and function. The greatest differences are those between prokaryotic and eukaryotic cells. Prokaryotes are generally smaller and simpler.

Eukaryotic cells are highly compartmentalized—most of their essential functions occur inside specialized structures enclosed by membranes. Like the organs of organisms such as multicellular plants and animals, each of these socalled "organelles" performs a particular task.

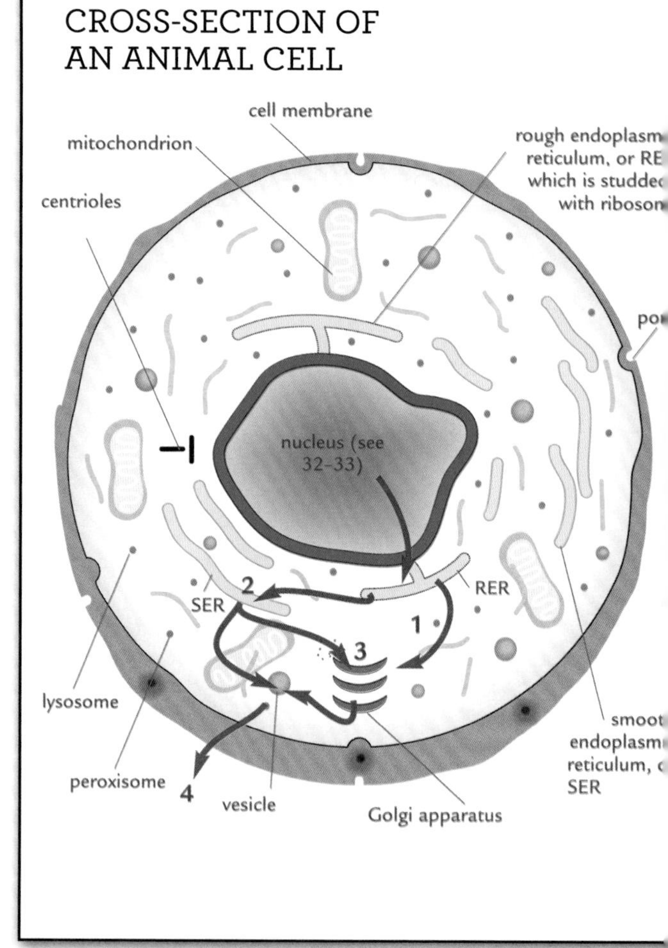

An animal cell. The nucleus controls all cell activities. It controls what proteins are made. Cytoplasm is the cell's contents outside the nucleus, including organelles. Most organelles are involved with making, processing, packaging, and transporting proteins. Mitochondria produce energy.

In prokaryotic cells there is much less organization: They contain deoxyribonucleic acid but no internal membranes. The only structures prokaryotic cells have in common with eukaryotic cells are the external cell membrane and internal structures called ribosomes. They have no organelles.

PROTOPLASM, CYTOPLASM, AND CYTOSOL

"Protoplasm" is a rather outdated word for the contents of a cell. Biologists now know that protoplasm is far from the uniform mass it was once described as. Cytoplasm is now used to describe the contents of the cell outside the nucleus,

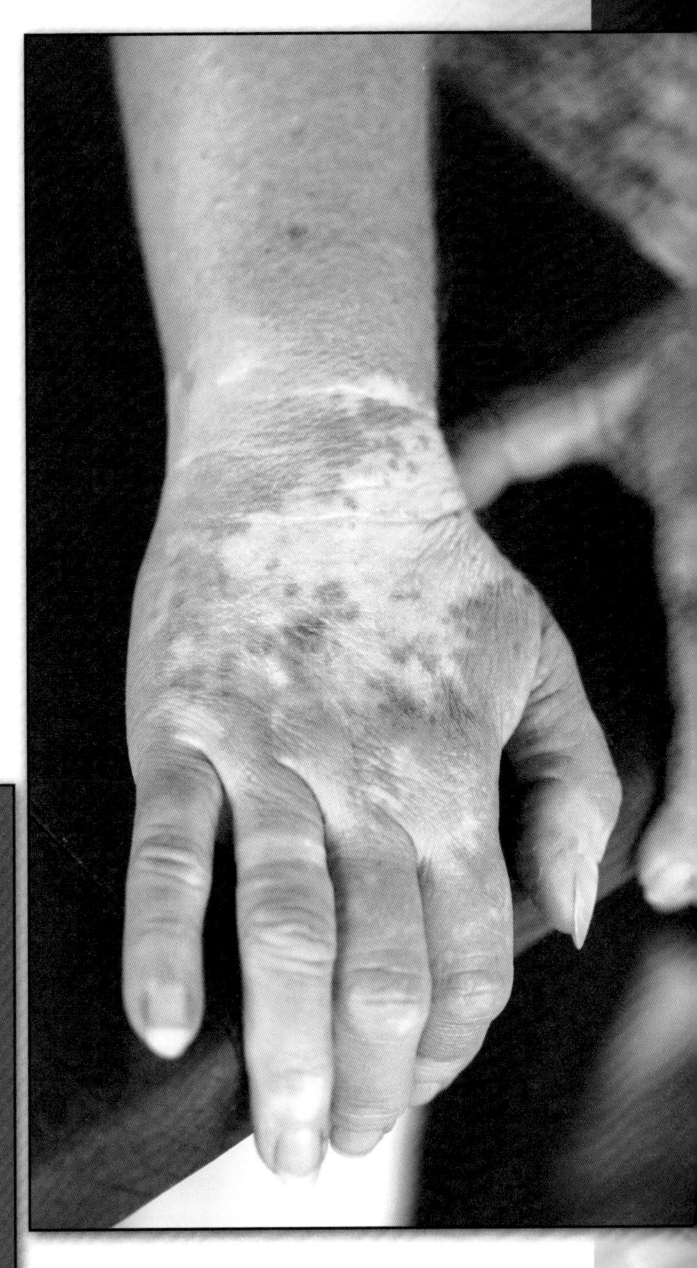

Skin cells called melanocytes contain pigments (molecules that produce color). Vitiligo occurs when melanocytes lose their ability to make pigments. Patches of the skin lose their color, and the normal skin color rarely returns. There is no cure.

CELL ORIGIN

The word "nucleolus" was first used by the German biologists Matthias Schleiden (1804–1881) and Theodor Schwann (1810–1882). They noted the similarities between plant and animals cells, and popularized cell theory, which states that living things are all made up of cells. Schleiden and Schwann believed that cells are made from a mass of structureless jelly called cytoblastema. The first stage in the development of a cell, they said, was the appearance of a tiny dark granule, the nucleolus. The nucleus and then the cell itself build up layer by layer around this tiny core. We now know that this idea was wrong, and that all cells form by division. The nucleolus, which does exist, has kept its name, though.

A LOOK AT THE ROUGH ENDOPLASMIC RETICULUM

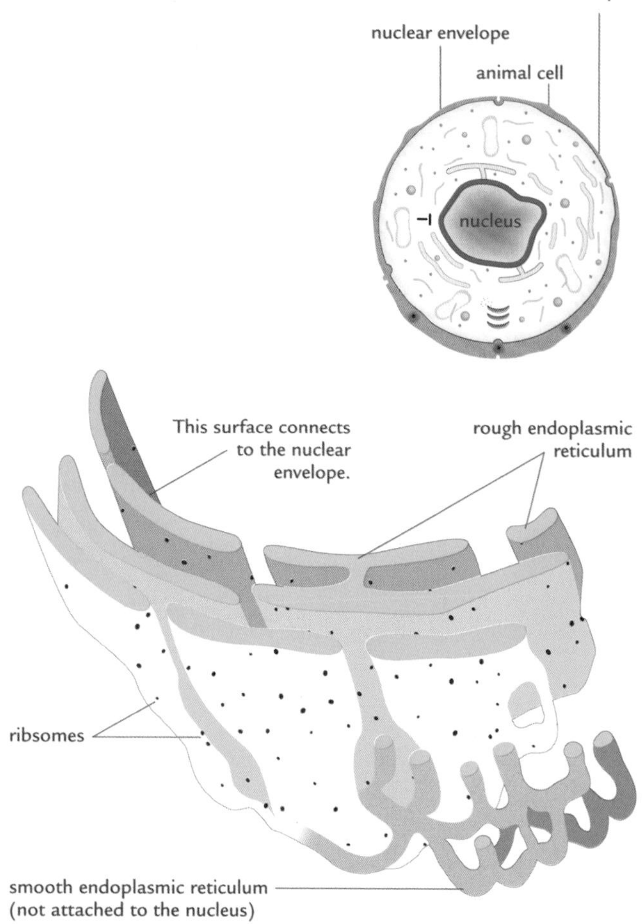

Rough endoplasmic reticulum (RER) is made up of many folded membranes. Ribosomes attached to the RER give it a rough surface. They are the sites for protein manufacture.

including organelles. Textbook diagrams of cells usually show the various internal structures or organelles floating in a jelly-like fluid, the cytosol, which also contains the cytoskeleton. The cytosol is a rich soup of chemical compounds. It is also the site of many cellular reactions. The cytoplasm itself is extremely complex. It incorporates the cytoskeleton, a supportive network of minute tubes and threads, and the cytosol as well as organelles.

UNDERSTANDING ORGANELLES

INSIDE THE NUCLEUS

The nucleus is the largest organelle in a eukaryotic cell. It can be seen under a relatively low-powered microscope, often without special treatment. The nucleus is bound by a double membrane called the nuclear envelope. The nuclear envelope is perforated by numerous openings or pores. They allow messenger molecules to move in and out of the nucleus. The messenger molecules carry instructions regarding the manufacture of essential cell components and molecules, such as enzymes. The nucleus of most cells also has an area called the nucleolus dedicated to the production of ribosomes.

In every body cell the nucleus contains a full copy of the organism's

MANY NUCLEI

Most cells have just one nucleus, but many single-celled protists are multinucleate—they have many nuclei. For example, amebas such as *Chaos* can have several nuclei. Ciliate protists like *Paramecium* have two types of nuclei, one large (the macronucleus) and one small (the micronucleus). Multinucleate cells also occur in some animal tissues, including certain muscle tissues.

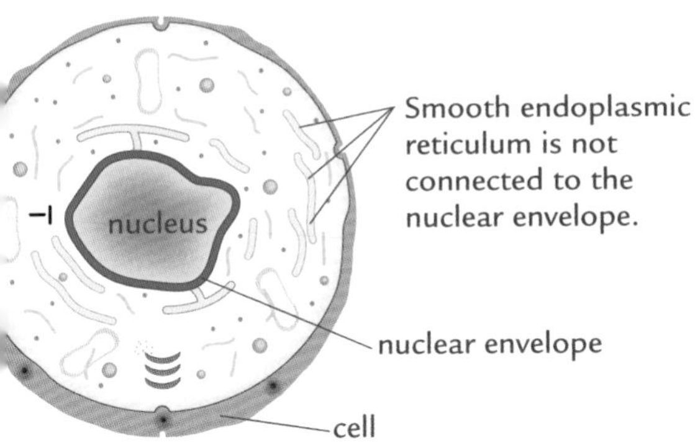

There are no ribosomes on smooth endoplasmic reticulum (SER). Proteins made on the RER are processed by the SER.

A CLOSER LOOK AT RIBOSOMES

Strictly speaking, ribosomes are not organelles because they are not bound by membranes. They are the only internal cell structure that occurs in both prokaryotic and eukaryotic cells. Ribosomes are either free in the cytosol or attached to the membranes of rough endoplasmic reticulum (which only occurs in eukaryote cells). Their job is to "read" the genetic instructions coded on strands of messenger RNA. They build the various proteins needed by the cell. Ribosomes are small—those of eukaryotic cells are about 30 nm in diameter, and prokaryotic ribosomes are even smaller.

genes (units of inherited information). Genes are segments of a molecule called deoxyribonucleic acid, or DNA. When a cell is not dividing, DNA forms tiny, invisible threads called chromatin. As a cell prepares to divide, the DNA copies itself. Then the chromatin coils to form chromosomes, which are divided between the two new daughter cells. Chromosomes unwind back into chromatin.

WHAT IS THE GOLGI APPARATUS?

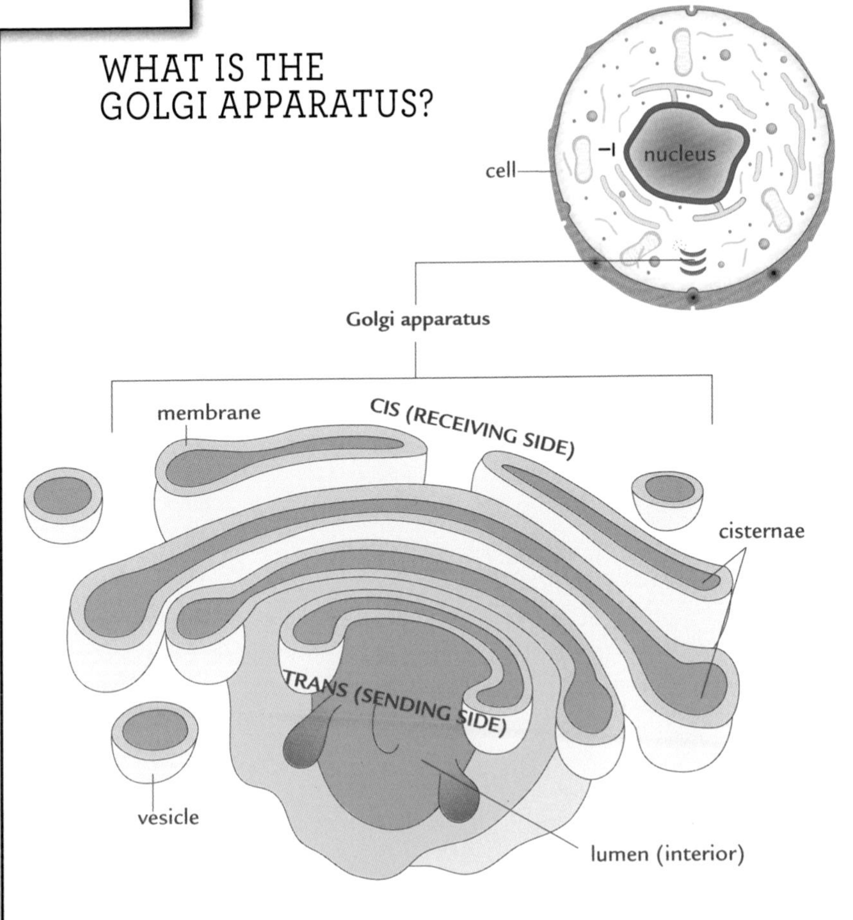

When a cell is not dividing, there is still plenty of activity inside the nucleus. For instance, the strands of DNA are used as templates to make messenger ribonucleic acid, or mRNA. In this way mRNA carries the genetic code on DNA from the nucleus to the rest of the cell. There the information on mRNA directs the production of proteins such as hormones and enzymes.

RER

The grandly named rough endoplasmic reticulum, or RER, surrounds the nucleus and is connected to the nuclear envelope. The RER's size varies, but it is largest in cells that manufacture large quantities of proteins. The word *reticulum* means "network" and refers to the complex shape of the RER. Endoplasmic describes the organelle's location in the cell (*endo* means "inside"). Rough refers to the fact that the outer membranes of RER are studded all over with tiny granular strucures called ribosomes.

RER is like a processing and manufacturing plant where the instructions coded on DNA in the nucleus are put into action. Ribosomes use strands of messenger RNA (made from DNA) to join amino acids into chains. These chains, called polypeptides, are made into proteins. The newly made proteins accumulate inside the RER. They are then packaged and transferred to the Golgi apparatus for further processing.

SER

Smooth endoplasmic reticulum, SER, has no ribosomes and is not attached to the nuclear envelope. SER has a tubular structure and forms stacks of flattened sacs, like piles of fat pancakes riddled with holes.

EXOCYTOSIS AND ENDOCYTOSIS

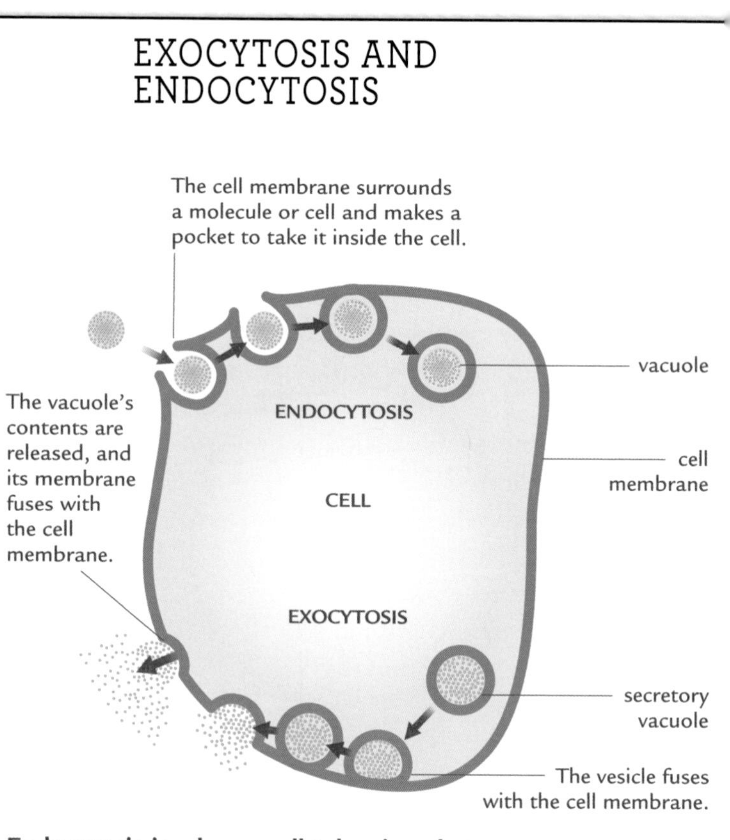

Endocytosis is when a cell takes in substances. Phagocytosis is a type of endocytosis involving particles, such as when white blood cells engulf foreign bodies. Pinocytosis is when a cell takes in liquids. Exocytosis is the process cells use to get rid of substances.

AN EXPERIMENT WITH CONTRACTILE VACUOLES

Use a microscope to examine a freshwater protozoan such as *Paramecium* mounted in a drop of fresh water under a coverslip. Look around the cell edge for a round vacuole that gradually increases in size, then suddenly disappears. It is a contractile vacuole (*Paramecium* have two). Using a stopwatch, time how long it takes the vacuole to expand and discharge 10 times, then figure out an average. Using a fine pipette, transfer the organism to a 2 percent solution of salt water (2 grams salt in 98 ml fresh water). Time 10 cycles. Repeat the process with 4, 6, and 8 percent salt solutions. Compare your results. Using water that contains more salt means there is less difference in concentration between the cell and its environment.

SER packages proteins that are due to be exported from the cell, such as secretory proteins. It is the site of synthesis for various lipids.

VACUOLES AND VESICLES

Also present in the cytoplasm of plant and animal cells is an assortment of membrane-bound organelles called vacuoles and vesicles. They are mainly used for storage or transport of food, waste, or various kinds of molecules made in the cell.

Vacuoles are generally temporary structures that form by budding off other membranes (in particular, the cell membrane, RER, SER, and Golgi apparatus). Vacuoles are reabsorbed into the membrane and recycled once their job is done.

UNDERSTANDING THE LYSOSOME CYCLE

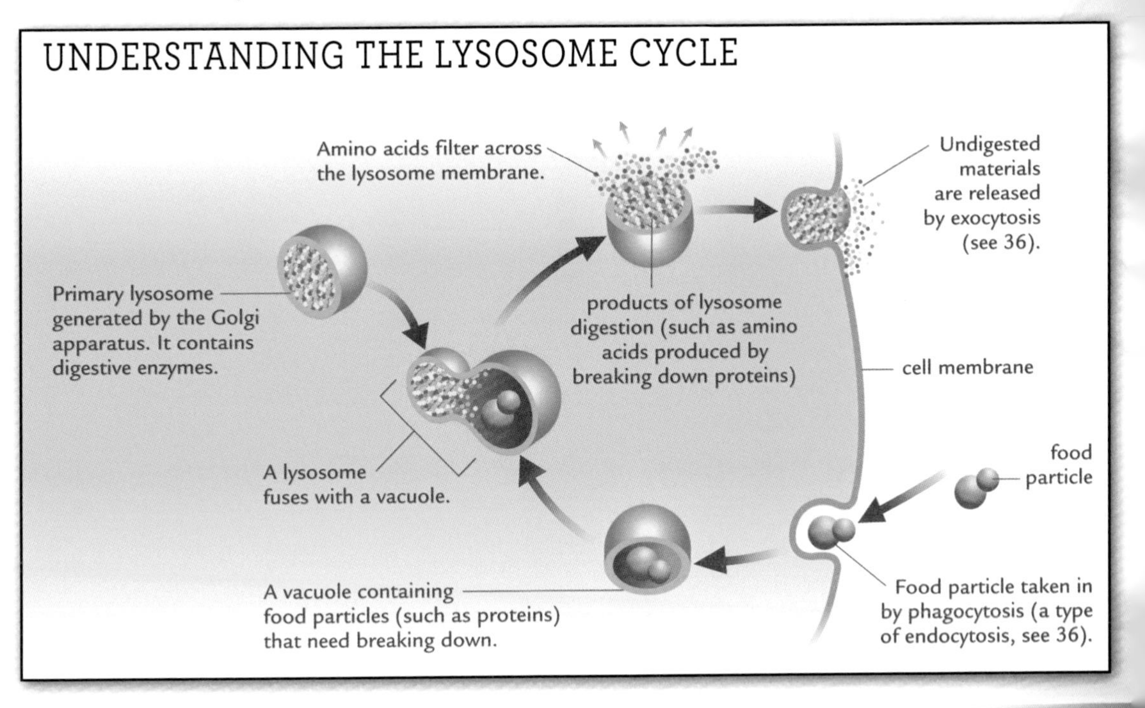

When vacuoles form from the cell membrane to bring some substance into the cell, the process is called endocytosis. The reverse process, by which vacuoles fuse with the cell membrane in order to eject their contents from the cell is exocytosis.

INSIDE THE CONTRACTILE VACUOLE

Aquatic protozoa, including ciliates, flagellates, and amebas, have a contractile vacuole. It acts like an organ for expelling excess water. The contents of a freshwater protozoan cell are more concentrated than the medium in which the organism lives. This means there is a constant tendency for water to enter the cell by osmosis. To prevent the cytoplasm from becoming dangerously diluted, or even exploding, the organism must actively expel water. The contractile vacuole absorbs water from the cytoplasm, expanding as it fills, then fuses with the cell membrane and expels water.

WHAT ARE LYSOSOMES?

Lysosomes are a feature of animal cells. They contain several different digestive enzymes that break down large molecules such as proteins, fats, and carbohydrates. Such molecules are taken into the cell by phagocytosis.

Lysosomes digest dead or unwanted material and appear to play an important

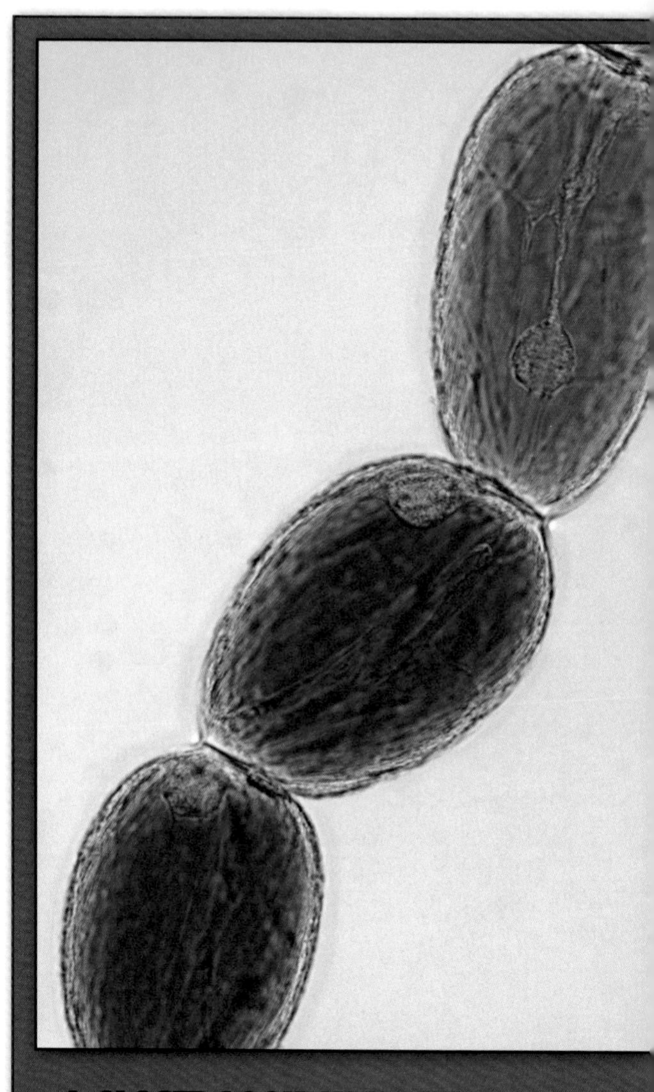

A CLOSER LOOK AT PLANT VACUOLES

One of the most distinctive features of plant cells, apart from the cell wall, is the vacuole, a large, apparently empty space in the middle. A vacuole contains mostly water, and it is important in keeping plant tissues rigid. As long as the plant is kept well watered, the vacuole remains inflated.

role in recycling materials when cells die. Lysosomes are usually 0.5-1.0 μm in diameter and are enveloped in a single layer of membrane. The digestive enzymes are manufactured on the rough endoplasmic reticulum, from where they are transferred to the Golgi apparatus. New lysosomes that have recently budded off the Golgi apparatus are called primary lysosomes. They become secondary lysosomes when they have fused with a vacuole and have begun breaking down the contents.

The products of lysosome digestion are tiny molecules such as amino acids. They are small enough to move across the lysosome membrane

ADENOSINE TRIPHOSPHATE AND A CHLOROPLAST

Chloroplasts capture the energy of sunlight and store it in molecules of the compound adenosine triphosphate (ATP). ATP is made in the thylakoid membranes of chloroplasts (1). ATP is used to make sugars in the stroma (2). Both these reactions make up the process of photosynthesis. When ATP is converted to adenosine diphosphate (ADP), it releases a burst of energy that fuels cellular activity. ADP changes back to ATP in respiration. In this process glucose and oxygen react to form water and carbon dioxide, along with the release of energy.

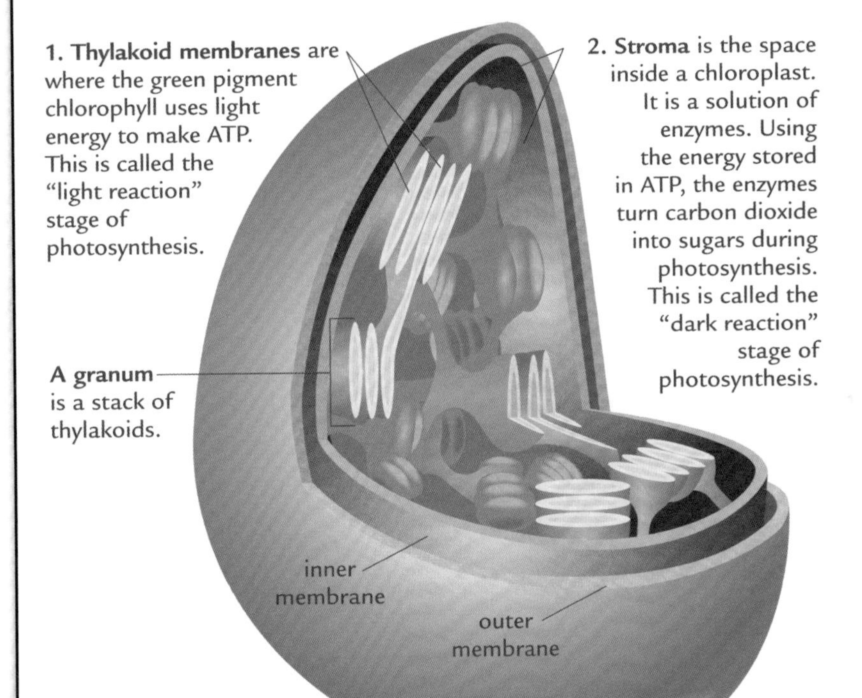

LYSOSOMES

Lysosomes were not discovered by microscopy but by biochemistry. In the 1950s cell biologist Christian de Duve (born 1917) was conducting experiments on the action of the hormone insulin. He noticed that some extracts of liver tissue were being digested by enzymes that appeared when the tissues were ground up. He thought that the enzymes must be leaking from mystery vesicles that ruptured during the preparation process. Later de Duve traced the source of the enzymes to a previously undescribed class of organelle, which he called lysosomes (from the Greek *lysis*, meaning "breakdown" or "digestion"). In 1974, de Duve won a Nobel Prize for his work.

and into the cytosol. Amino acids are then used to make proteins. Any indigestible material still in the lysosome is eventually expelled from the cell when the organelle fuses with the cell membrane.

PEROXISOMES

Peroxisomes are small organelles (0.5–1.0 μm) enclosed by membranes. Peroxisomes are similar to lysosomes and occur in both plant and animal cells. They have several functions, including packaging and breaking down poisons such as hydrogen peroxide, methanol, and other potentially harmful substances. In animal cells peroxisomes assist in the chemical changes of certain fatty acids that cannot be effectively processed by other organelles. In plant cells peroxisomes are usually associated with mitochondria and chloroplasts. They are also involved in chemical changes, including converting stored fats into useful carbohydrates.

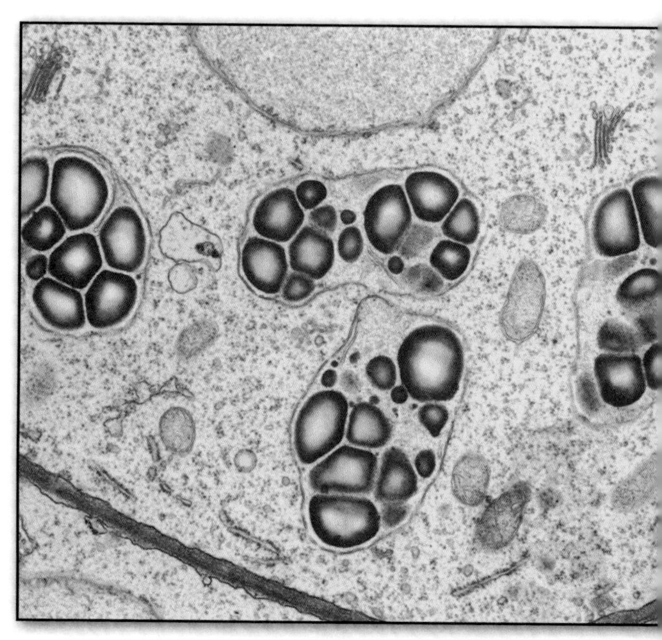

Leucoplasts are organelles in plant cells. They act as stores for starch granules.

WHAT ARE CHLOROPLASTS?

The term *chloroplast* means "green particle." These prominent organelles give green plants their color. Chloroplasts belong to a group of organelles generally called plastids. In plants such as copper beeches, red sycamores, and in red and brown seaweeds the plastids contain other pigments that mask the underlying green color. These plastids are called chromoplasts (colored particles). Chloroplasts are

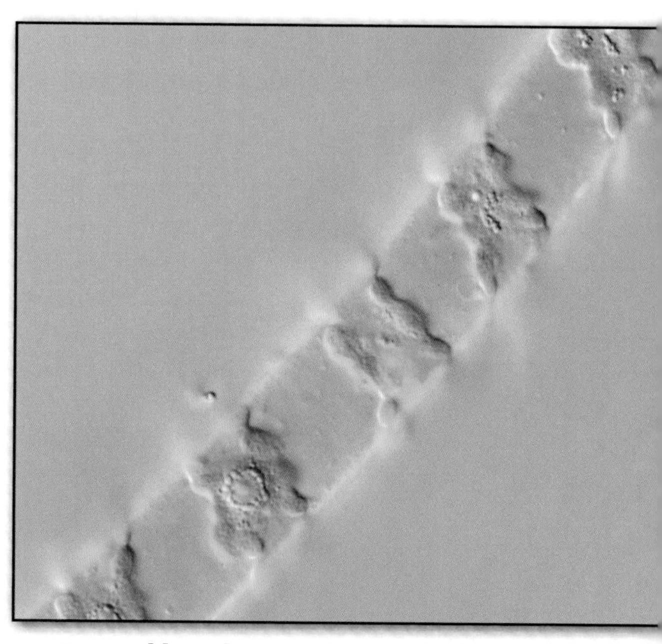

Green chloroplasts can be seen in this strand of green alga (*Sprirogyra*).

A WOMAN KNOWN AS MITOCHONDRIAL EVE

Mitochondria contain small amounts of DNA that replicate (copy themselves) independently of DNA in the nucleus. Mitochondrial DNA (mDNA) is unique because plants and animals inherit it all from their female parent. Differences in mDNA sequences between generations can only occur by random mutations (changes). Such mutations are rare, but reasonably constant. It has thus been possible to measure the amount of differences in mDNA. This provides a kind of molecular clock for estimating the number of generations and therefore the time that must have elapsed since two or more individuals had a common (female) ancestor. The technique has been used to suggest that the most recent common ancestor (through the female line) of all humans alive on Earth today was a woman who lived in Africa about 200,000 years ago. This woman is known as Mitochondrial Eve.

large—often up to 10 μm long. They occur in all photosynthetic cells, from tiny single-celled algae such as *Chlamydomonas* to the cells that form the leaves of tall trees. There may be just one or many chloroplasts per cell. Each is enclosed in a double membrane layer and has a complex arrangement of flattened sacs called thylakoids. Thylakoids contain chlorophyll. They are arranged in interconnected stacks called grana (sing. *granum*) and are surrounded by a fluid called the stroma.

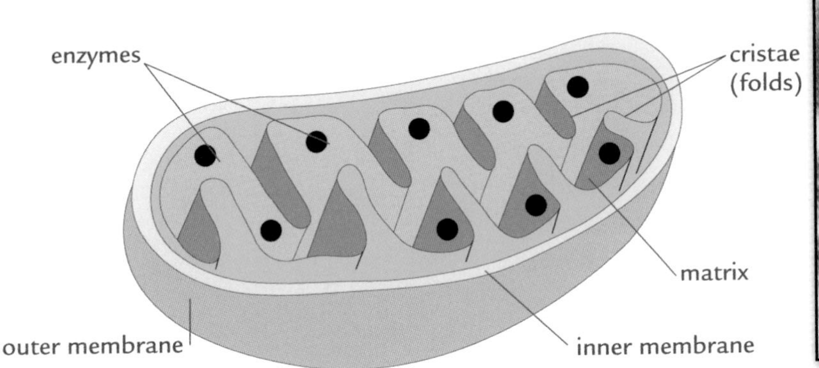

STRUCTURE OF A MITOCHONDRION

The inner membrane of a mitochondrion forms folds called cristae. Respiration—the process of energy production using glucose and oxygen—occurs on the cristae.

GAINING ENERGY FROM FOOD

You will need: unsalted peanuts, a coffee pot holder (or large can), a lighter or matches, a small ringpull can, foil, scissors, a jar with a lid, a cooking thermometer, water, tape.

1

1 Wrap foil around the coffee-pot holder on either side of the handle. Tape the edges together to form a tube. Fold the ends of the tube neatly. Cut two slots in the foil near the top of the pot.

Be very careful if you use matches or a lighter. Always ask an adult to help you.

2 Fill a clean, empty ringpull can with water. Measure its temperature with a cooking thermometer. Now place a peanut on the lid of the small jar.

3 With the help of an adult, light the peanut with a match or lighter. Then slot the coffee-pot holder or large can over the lit peanut with the base uppermost. Now rest the small can of water on top, on the base of the coffee-pot holder.

Once the peanut burns out, use the thermometer to test the temperature of the water in the can. The water should now be much warmer, showing the peanut has given out real heat energy.

3

It is impossible to overstate the importance of chloroplasts to life on Earth. They are the site of photosynthesis, the process by which plants use sunlight to make carbohydrates (the chemical fuel used by all cells) from carbon dioxide and water. Photosynthesis is the basis of the food chain in which plants become food for animals. In addition to photosynthesis chloroplasts often also have a role in storing starch granules or lipid droplets. Some plants have plastids called leucoplasts, which store starch granules.

A LOOK AT MITOCHONDRIA

After the nucleus, mitochondria (sing. *mitochondrion*) are usually the next most conspicuous features of animal

Food provides energy for animals. They need it to keep warm and for movement. Energy release occurs by a chemical change when cells "burn" food fuel. Foods like peanuts are high in energy.

and fungal cells. Other eukaryotes such as plants and seaweeds have them, too, but they are often dwarfed by chloroplasts. Mitochondria are large—usually a few microns in length, about the size of a bacterial cell. Their structure is very distinctive. They are bound by a double-layered membrane. The outer layer of the membrane is relatively smooth, enclosing the organelle in a fairly regular oval shape. The inner membrane is highly folded and forms a series of layers called cristae.

Mitochondria are the cell's power plants. Inside them cellular fuels such as sugars and fats are converted into useful chemical energy. The most important product of the process is the energy-storage compound adenosine triphosphate (ATP). ATP is a molecule that readily converts to adenosine diphosphate (ADP). Simultaneously it releases a burst of heat energy. ATP produced by mitochondria is collected by other cell components that need energy to perform their various functions.

While exercising, cells in the human body burn up food. A chemical change occurs during which energy and warmth are released. Sweating helps cool the body down again.

INSIDE THE CENTROSOME

Eukaryotic cells except flowering plants, pine trees, and some protists contain a structure called the centrosome. It is positioned near the nucleus It contains a pair of rodlike structures called centrioles. In turn the centrioles are made of microtubules. In cell division centrioles have a role in the formation of the mitotic spindle, which chromosomes are moved around by.

ADREOLEUKODYSTROPHY

The sex-linked disease adreoleukodystrophy, or ALD, affects young boys. It was brought to the world's attention by the 1992 movie *Lorenzo's Oil*. It showed the efforts of Augusto and Michaela Odone to find a treatment for their son Lorenzo, who was dying from ALD. ALD is a disease caused by the failure of peroxisomes to chemically change certain fatty acids. The treatment Lorenzo's parents discovered was another fatty acid, now called Lorenzo's oil, that may stop the progress of the disease when taken as part of a strictly controlled low-fat diet. The treatment is still controversial because scientists are still not sure precisely how or if it works.

CHAPTER FIVE

HOW CELLS COMMUNICATE

Communication inside and between cells is vital for the smooth running of an organism's various body systems.

Your body is made up of cells that form tissues. They are groups of cooperating cells that share a common function. Tissues make up organs, such as the stomach, liver, or kidneys. For a cell to function correctly, different parts of it need to communicate. Similarly, all the cells within an organ must be in contact with each other so they can work together. Cells also need to send and receive information to and from organs such as the brain. Signals can be of two types. Some are chemical messages that trigger responses in other cells. Other messages take the form of electrical impulses.

Cell communication is essential for the well-being of the body. Sometimes communication breaks down, as in the disease diabetes. Diabetics must check their blood-sugar levels regularly. Here a diabetic tests for sugar in blood drawn from a prick in the finger.

CELL ADHESION MOLECULE

In multicellular (many-celled) creatures such as people it is vital that cells in a tissue can recognize each other. Cell surfaces contain a chemical called a cell adhesion molecule, or CAM. The CAM allows the cell to stick only to similar cell types. For example, a CAM enables liver cells to attach only to other liver cells. That keeps all the cells in a tissue organized and prevents other cell types from getting in the way.

JUNCTIONS

Adjacent cells are linked by several types of cell junctions. Tight junctions occur in the layers of cells that surround body cavities like the gut. Formed by the binding of proteins that extend out from inside the cells, tight junctions prevent even tiny molecules from passing between the cells. So a molecule moving from the gut of an animal into its bloodstream must go into the cells of the gut wall first. Tight junctions stop the molecule from slipping through the spaces in between.

STICKING TOGETHER

It is well-known that if a sponge is mashed up and run through a sieve, the cells will join again to re-form a complete sponge. How do sponges manage this neat trick? Sponge cells have molecules (called CAMs) of a protein protruding from their surfaces. When two proteins come into contact, they bind, joining the cells. That allows the sponge to re-form.

Sponges have no organs. But similar proteins allow cells to form tissues and organs inside the bodies of other animals.

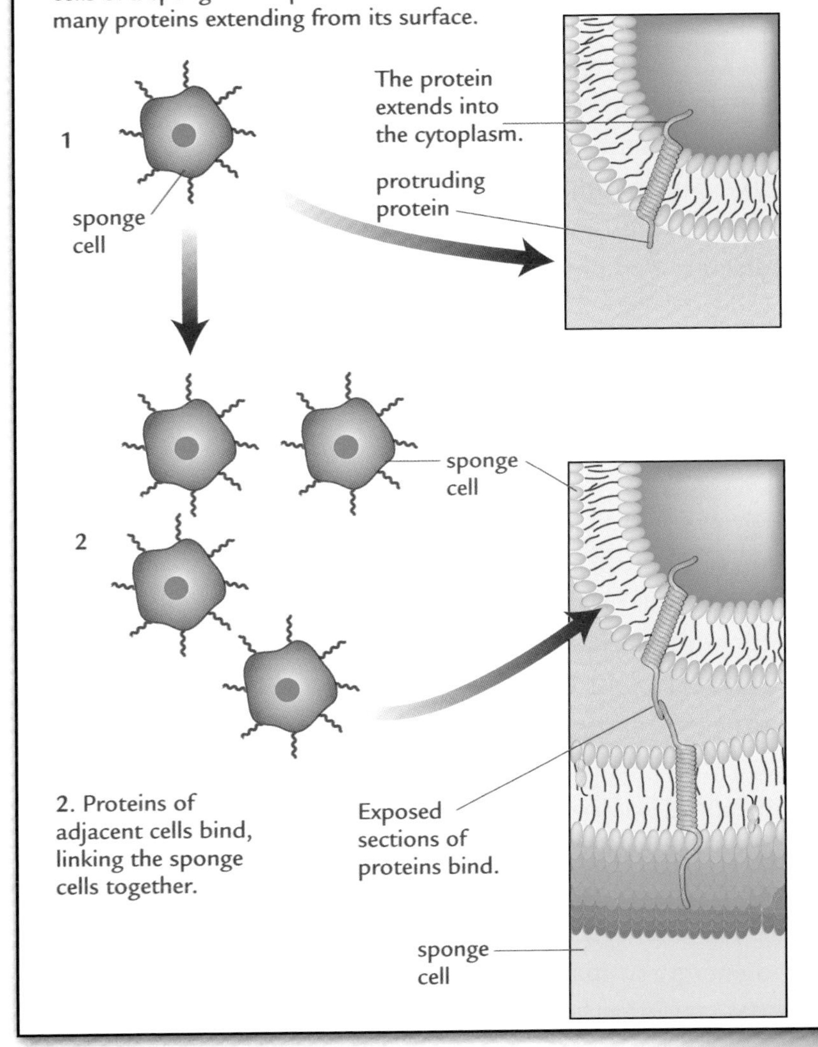

1. After being passed through a sieve, the cells of a sponge are separated. Each has many proteins extending from its surface.

2. Proteins of adjacent cells bind, linking the sponge cells together.

COMMUNICATION

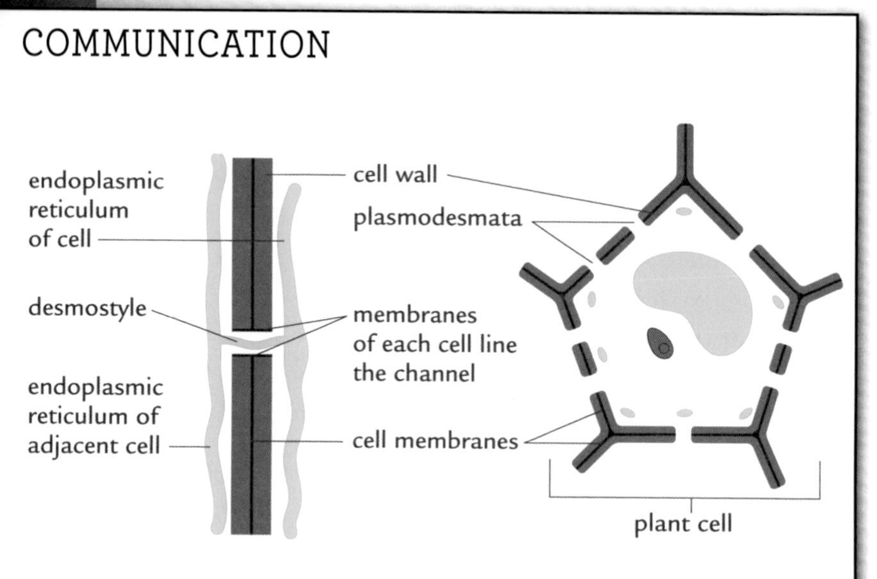

Instead of gap junctions, plants have structures called plasmodesmata. They link the cytoplasms of adjacent cells so water, ions, and small molecules can pass through easily. A tube called a desmostyle runs through each channel. This is continuous with the endoplasmic reticulum of each of the adjacent cells.

Desmosomes are joints that hold cells together firmly in other parts of the body. These joints are extra tough. Desmosomes occur in places of wear and tear, such as between skin cells. But for cells to communicate, another type of junction, called a gap junction, is needed.

WHAT ARE GAP JUNCTIONS?

A gap junction is a minute channel made of six protein tubes called connexons. It runs between

HEART ATTACKS

Like other cells, heart cells contain organelles called mitochondria that produce energy. Scientists have found that heart cell death during a heart attack can be reduced by keeping open gap junctions (right) in the mitochondria. During a heart attack heart cells are deprived of oxygen and energy, so many cells die. A junction in the mitochondria allows the movement of certain ions. By keeping these channels open, the injured heart can maintain energy and keep far more of its cells alive. Scientists are hoping to develop a molecule that will keep these gap junctions open. The treatment could be used just after a heart attack to prevent further cell death.

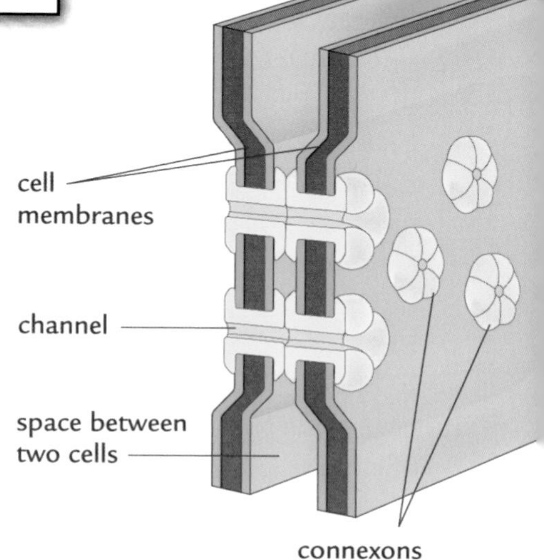

the walls of adjacent cells. Small molecules such as salts and sugars move between neighboring cells through these channels. Gap junctions are vital for tissue function. For example, muscle cells contract in unison because electrical signals pass from one muscle cell to the next. The electrical signals are caused by the movement of particles called ions. Ions are small enough to move through gap junctions.

Gap junctions are not open doors; they do not transmit just any signal. They can change their shape, and thus the types of chemicals that can get through, depending on the needs of the cell.

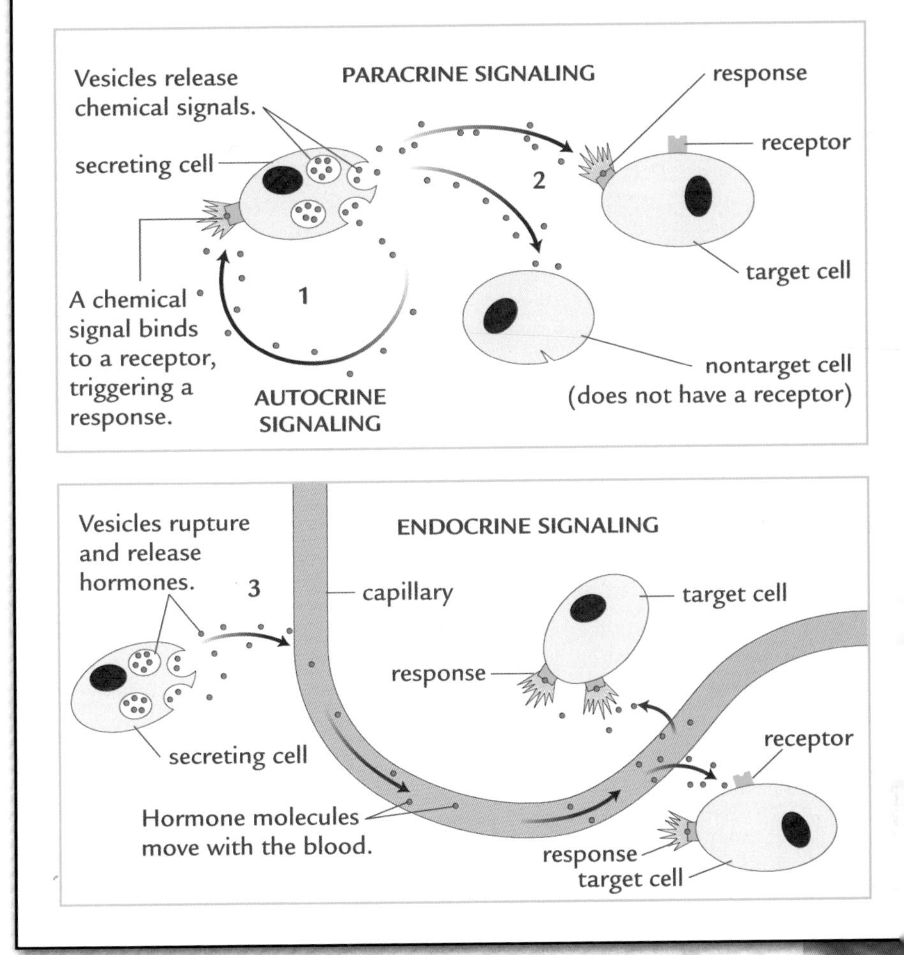

There are three types of chemical signals. Autocrine signals (1) involve one part of the cell releasing chemicals that trigger a response in another part of the same cell. Paracrine signals (2) are targeted at nearby cells, while endocrine signals (or hormones) act over longer distances (3).

Most gap junctions transmit electrical ion signals, although there are very many exceptions. Gap junctions in nonelectrically signaling tissues allow the transport of nutrients and waste in and out of cells, for example.

WHAT ARE CHEMICAL SIGNALS?

There is a lot going on inside a cell. To ensure that everything functions as it should, different parts of a single cell need to communicate with each other. Chemicals in the cytoplasm function as signals, telling the cell what to do. That is called

SUGAR LEVELS

There are many examples of the importance of communication in the body and the dangers of cutting the lines of contact. The level of glucose, a sugar, in the blood is kept under strict control and must be maintained within a very narrow range. The pancreas is an endocrine gland. It secretes hormones called insulin and glucagon that regulate blood-sugar levels. The pancreas gets feedback from the body that allows it to regulate the amounts of these hormones it releases. If there is too much sugar in the blood, the pancreas secretes insulin, which helps cells absorb the excess. If there is too little sugar, glucagon is released. Glucagon molecules communicate with the liver, stimulating it to release glucose from storage. Many levels of cell communication are involved in maintaining blood-sugar levels. If communication between the blood and the pancreas is disrupted, as in the disease diabetes, blood-sugar levels can soon drift outside normal ranges, which can be deadly.

intracellular signaling. Chemicals can be released and responded to by the same cell. That is called autocrine signaling.

Cells also use chemicals to communicate with other cells over short distances. This is called paracrine signaling. Chemicals used in paracrine signaling are often produced in high concentrations.

Chemical communication between cells over longer distances in the body is called endocrine signaling. Endocrine signals are formed by hormones. They are chemicals secreted by endocrine glands, which include the pituitary, adrenal, and thyroid glands. The hormones are carried through the blood to the places where they are needed. Hormones are among the

THE FUNCTION OF PAINKILLERS

Painkillers (right) block the chemicals that pass the sensation of pain through the nervous system to the brain. There are two types of painkillers. One type, including aspirin, halts the body's production of prostaglandins, chemicals that produce the sensation of pain. Other painkillers block receptors in the brain so the pain sensations cannot be received.

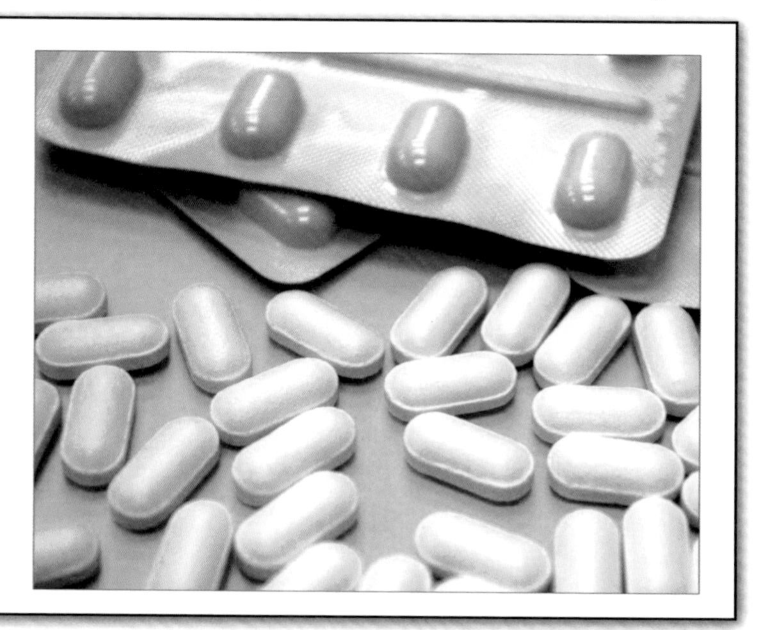

most common of the body's chemical signals, although each is usually produced in low concentrations.

CHEMICAL SIGNALS

All cells have receptors on their surfaces. A receptors' shape only allows it to bind with specific molecules. The molecule (such as a hormone) that binds to a receptor is its ligand. The receptors are made of proteins, which are intricately folded molecules. The number, type, and arrangement of proteins determine which of the chemical signals carried in the blood the receptor is able to receive.

The receptor–ligand system is like a lock and key: The receptor is the lock, while the ligand is the only key that can open it. When the lock is opened, the ligand can get to work on the cell. Once bound with a ligand, a cell receptor responds in one of several ways. Some receptors move into the cytoplasm with the ligand attached. Other receptors activate molecules in the cell membrane that create new chemicals. The new chemicals carry the message into the cell. Some receptors create channels in the

INTRODUCING IONS

Electrical signals in the body depend on the movement of particles called ions. All matter is made up of tiny particles called atoms. Each atom has a nucleus orbited by one or more tiny electrons. The nucleus contains positively charged particles called protons and usually also particles called neutrons that do not have a charge. Electrons have a negative charge. In an atom the positive and negative charges balance each other out. However, an atom may lose or gain one or more electrons. It then becomes electrically charged. These charged atoms are called ions.

For example, table salt is a chemical compound called sodium chloride, or NaCl. When dissolved in water, the compound separates out, producing two types of ions. Sodium ions lose one electron, so they are positively charged (Na^+). Chloride ions hang onto the electron lost by the sodium, so they are negatively charged (Cl^-).

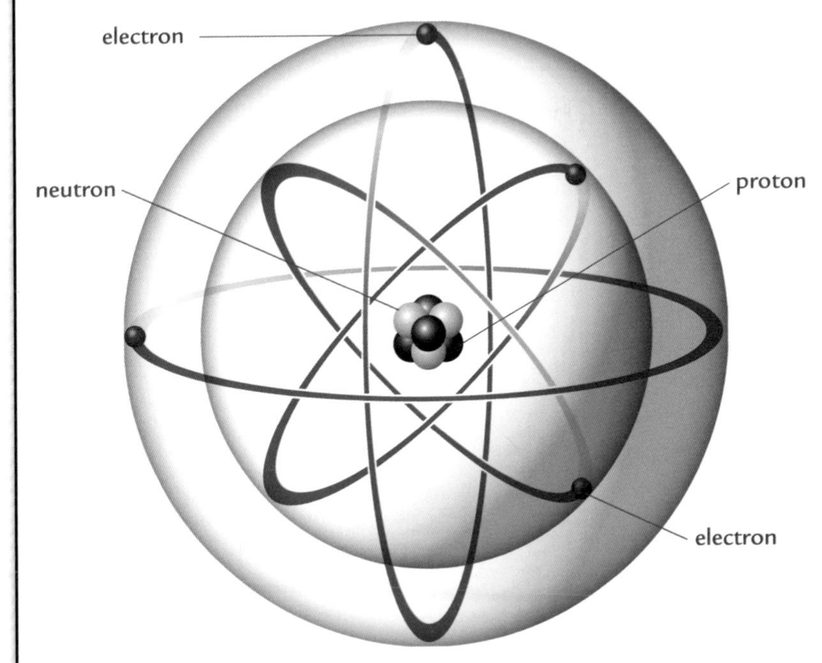

This is an atom of the metal beryllium (Be). This element forms ions by losing its two outer electrons to become Be^{2+}.

cell membrane through which ions can move. The movement of ions across the cell membrane alters the cell's electrical properties.

Whichever type of action takes place, the binding of ligand molecule to receptor causes a cascade of messenger chemicals to move inside the cell. These messengers carry out the tasks directed by the ligand.

IMPORTANT CAMP

One of the most important messengers inside the cell is a molecule called cyclic AMP, or cAMP. Normally the concentration of this messenger in a cell is very low. But when a ligand binds to a receptor, an enzyme in the cell membrane starts churning out lots of cAMP. The cAMP moves from the cell membrane into the cytoplasm. There it activates a range of enzymes. They are proteins that help chemical reactions take place and so control the behavior of the cell. The cAMP can also trigger a cascade of chemical reactions that speed up the cell's response to the ligand.

Sometimes such chemical go-betweens are not required. In these cases the receptor– ligand complex causes parts of the cell membrane to bunch up. The bunched-up segment of membrane moves into the cell. There it fuses with certain organelles inside, activating them. The receptors on cells triggered by insulin and other hormones are called tyrosine kinase receptors. Sometimes these receptors malfunction, and communication with the cell breaks down. Problems with tyrosine

The greatest heavyweight boxer of all time, Muhammad Ali (born 1942). After his retirement from boxing in the 1980s Ali's voice began to slur, he developed uncontrollable shaking, and he had trouble with limb coordination. These problems were symptoms of Parkinson's disease, of which Ali is a sufferer.

PARKINSON'S DISEASE

Parkinson's disease is caused by a deficiency of a chemical called dopamine. It is caused by the death of cells that make this substance. Amounts of another brain chemical, acetylcholine, then increase. It is this chemical that causes the tremors that characterize Parkinson's. The disease can be treated with a dopamine substitute called L-dopa. Although the drug eases the symptoms, it cannot cure the disease.

HOW SYNAPSES WORK

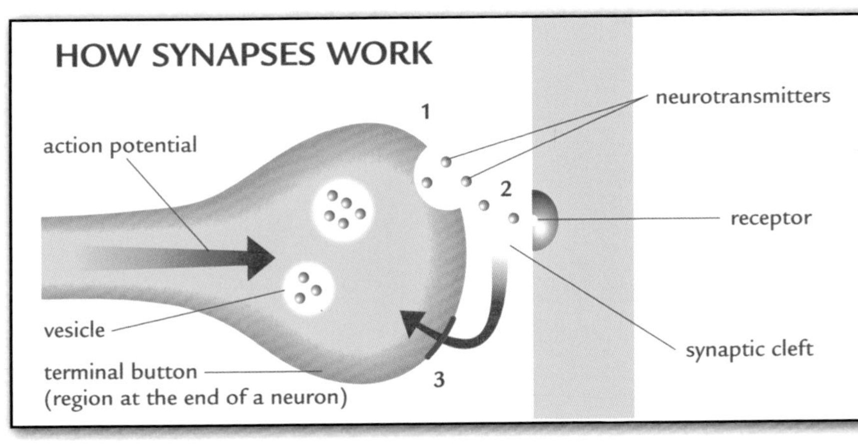

An action potential triggers the rupture of packets of neurotransmitters (**1**). The chemicals pass through the synaptic cleft. They bind to receptors on another neuron (**2**), causing another action potential. Spare neurotransmitters are then reabsorbed (**3**).

kinase receptors may lead to diseases like type II diabetes.

COMMUNICATING WITH ELECTRICAL IMPULSES

It is not just chemicals that carry messages around the body. Cells also communicate by using electrical impulses. This forms the basis of the nervous system. Nerve cells, or neurons, carry impulses between your brain and every other part of your body.

Neurons are master communicators, both with each other and with the cells that form the tissues and organs of the body. All neurons contain dendrites, thread-like extensions that receive messages from other nerve cells and transmit the messages to the neuron cell body. The cell body contains the nucleus, mitochondria, and various other organelles. The axon is a long, stringy extension of the neuron. It conducts messages from the nerve cell body to other neurons.

The messages carried by neurons are electrical. Each message is called an action potential. It is a spike of electrical activity caused by the movement of different types of ions in and out of the neuron.

Messages move between neurons across a junction called a synapse. The journey requires tiny molecules called neurotransmitters. They cross the synapse and bind to receptors on the next neuron along. That prompts a new action potential, allowing the electrical message to go on.

A CLOSER LOOK AT MULTIPLE SCLEROSIS

Myelin is a fibrous sheath that surrounds nerve cells in the brain, spinal cord, and optic nerves. Myelin also plays an important role in helping neurons transmit electrical signals around the body. Without the myelin sheath nerve cell communication is severely hampered. Multiple sclerosis (MS) is a disease in which the body's immune (self-defense) system attacks the myelin. That leads to problems with eyesight, chronic fatigue, the loss of limb coordination, and organ failure. Scientists do not know exactly what causes MS, but its interruption of signal transmission along neurons can be fatal.

CHAPTER SIX

UNDERSTANDING THE CELL CYCLE

Most cells go through a series of life stages called the cell cycle.

Nearly all cells, whether they are free-living single-celled organisms such as bacteria or part of the tissues of a multicellular organism, go through a series of life stages called the cell cycle. The cell cycle includes cell growth, copying of its genetic information, or DNA, and finally, the division of one cell into two.

THE DIFFERENT CYCLES

There are two types of cells. They are called prokaryote and eukaryote cells. Prokaryotes are an ancient group of single-celled organisms that include bacteria. Prokaryote cells do not have nuclei. Instead, their single strand of DNA floats freely within the cell cytoplasm.

CELL DEATH

Some cells opt out of the cell cycle, but they do so at a cost—death. However, death can be an inevitable part of a cell's function. Take toenails, for example. The cells that make up your toenails and fingernails become filled with a tough, flexible protein called keratin. The keratin swiftly kills the cells, which are more useful dead than alive. The tough, dead nail cells protect the sensitive upper parts of toes and fingers from wear.

WHAT IS BINARY FISSION?

Binary fission is a form of asexual reproduction. Genetic material is not exchanged, so bacteria have little genetic diversity. However, binary fission does allow bacteria to increase their numbers with amazing swiftness. Bacteria still shuffle their genes occasionally, through a process called conjugation.

Animals, plants, fungi, and protists all consist solely of eukaryote cells. The DNA of eukaryotes is parceled up inside the nucleus. Eukaryote cells also contain miniorgans called organelles, which prokaryotes lack. Organelles include the chloroplasts and the mitochondria.

When not dividing, cells function in similar ways. They take in food, convert it into energy, and manufacture proteins and other products. However, there are some distinct differences in the ways prokaryote and eukaryote cells divide in two.

UNDERSTANDING BINARY FISSION

Prokaryote cell division is called binary fission. Before the cell divides, its single DNA strand replicates, or self-copies. The two DNA copies attach to the inside of the cell membrane. The membrane begins to stretch. As it elongates, the DNA strands are pulled apart. Soon the cell is so stretched that it has effectively doubled its size. Then the cell membrane begins to pinch inward at the center of

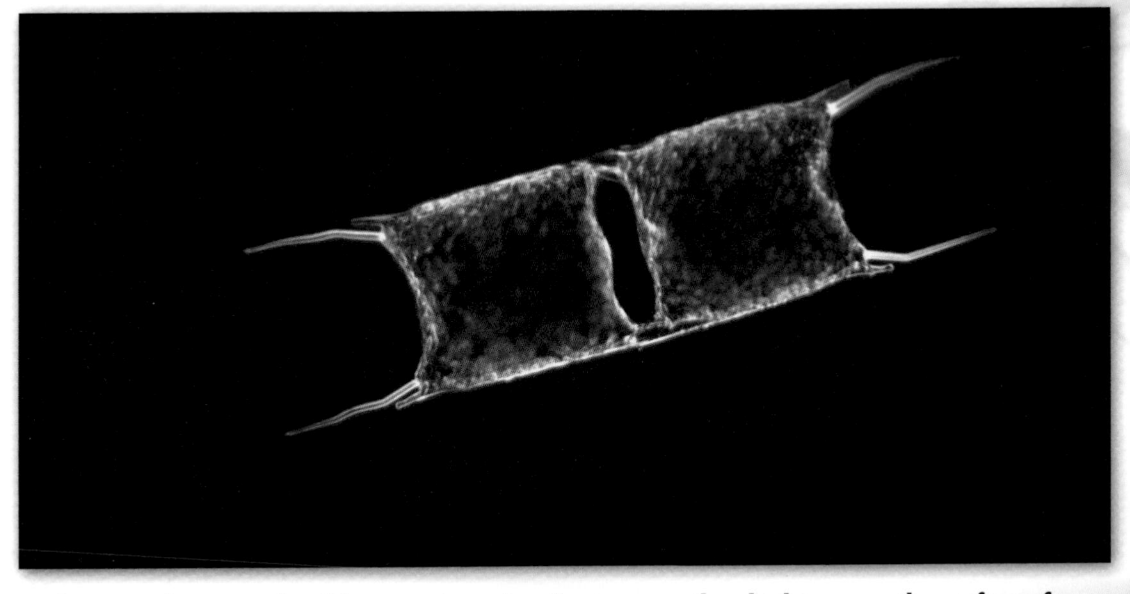

These are diatoms, plantlike organisms that float among the plankton near the surface of the ocean. Binary fission of one diatom into two has just come to an end.

the long, elongated cell. The cell's "waist" gets smaller and smaller as the membrane pinches inward. Eventually the two parts of the membrane meet in the middle of the cell.

Once the cell is completely pinched in two, a new cell wall forms in its center. The two new (or daughter) cells then split apart. The cells are genetically identical. One has the original DNA from the parent cell, while the other has an exact copy.

UNDERSTANDING THE EUKARYOTE CELL CYCLE

Nearly all eukaryote cells go through the same stages, or phases, during their cell cycle. There are four phases, each designated by a letter. They are the G1 phase, S phase, G2 phase, and M phase. The G stands for gap, the S stands for synthesis, and the M stands for mitosis.

DURING INTERPHASE

Phases G1, S, and G2 are together known as interphase. During interphase the cell uses energy to perform functions like the production of proteins. At the end of interphase chemical signals within the cell

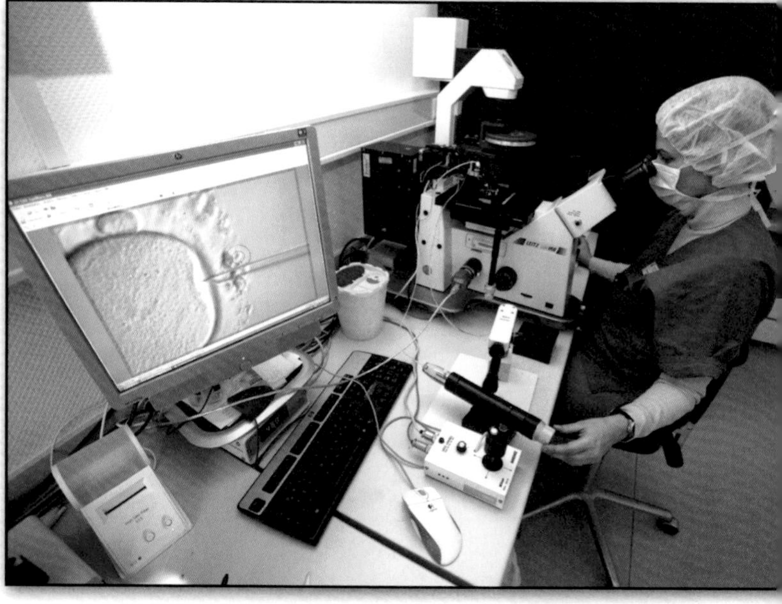

In vitro fertilization in action. The physician is carefully injecting sperm into an egg.

IN VITRO FERTILIZATION

Cells in a culture can move through their cell cycles very quickly. Medical researchers exploit this to produce drugs. They implant genes that produce drugs into cells. The cells increase in number until the drug can be harvested. Couples that struggle to have children sometimes have in vitro fertilization (IVF) treatment. Eggs are fertilized with sperm outside the body. The fertilized eggs are then placed in the woman's uterus. To make the woman's ovary release the extra eggs needed, a hormone, FSH, is required. FSH was once drawn from othe urine of women going through menopause. Today, human genes that code for FSH are implanted into cells from the ovary of a Chinese hamster. The cells are cultured, and the hormone is collected. This process produces larger quantities of FSH than existing techniques and also eliminates the possibility of infection.

INTERPHASE

In the nineteenth century biologists only had light microscopes to observe cells. They could see absolutely nothing of interest going on inside a cell during interphase. So interphase was dubbed the cells' resting phase. Biologists now know that a cell gets no rest at all during this time. Through electron microscopy biologists have shown that the cell seethes with activity during interphase. DNA replicates in the nucleus, while the organelles work in a frenzy, producing proteins and other products as well as liberating energy from food.

prepare it for cell division. The DNA of the cell must replicate, and a supply of organelles for the daughter cells must also be produced.

During the first gap phase, or G1, the cell increases the amount of cytoplasm it contains. It doubles in size and builds many of the extra organelles. G1 is usually the longest phase in the cell cycle. Chemicals determine how long G1 lasts. When the cell stops growing, G1 comes to a halt. More chemicals then launch the cell into the next stage, the S phase. The S, or synthesis, phase involves the replication of the cell's DNA inside the nucleus. The proteins needed by the cell for division are also created during this phase. Once DNA replication is complete, the cell moves into the second gap phase, G2.

During G2 the cell begins to put together all the structures it will need to separate its genetic material, allowing it to produce two daughter cells.

THE PHASES OF THE CELL CYCLE

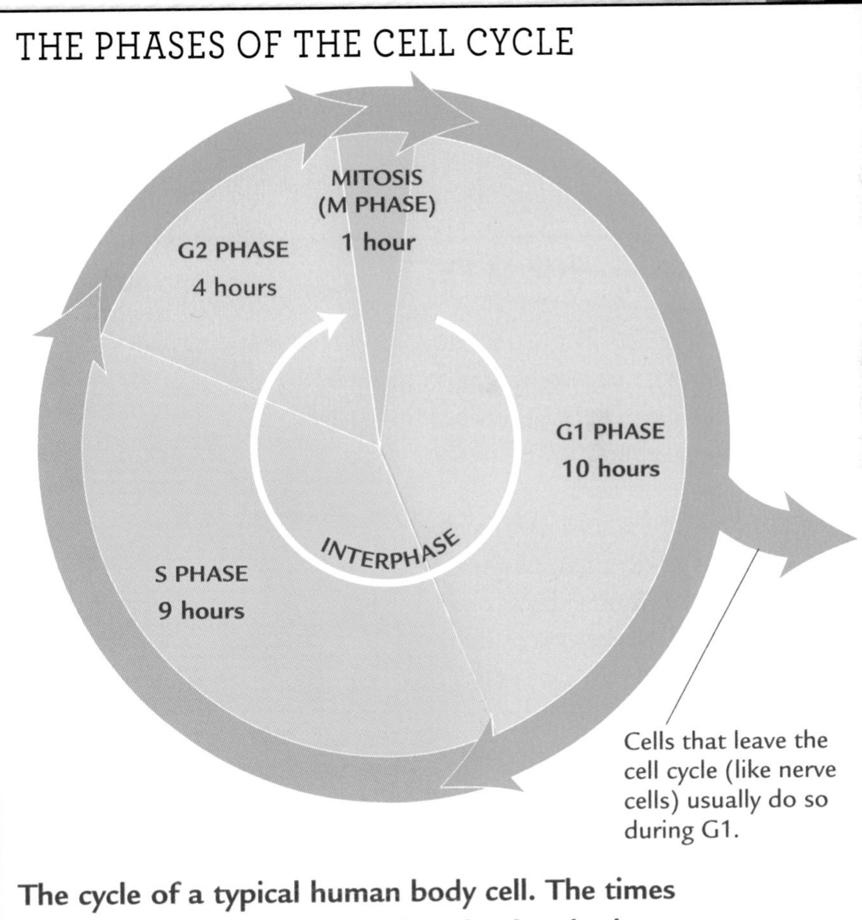

The cycle of a typical human body cell. The times shown represent the typical length of each phase.

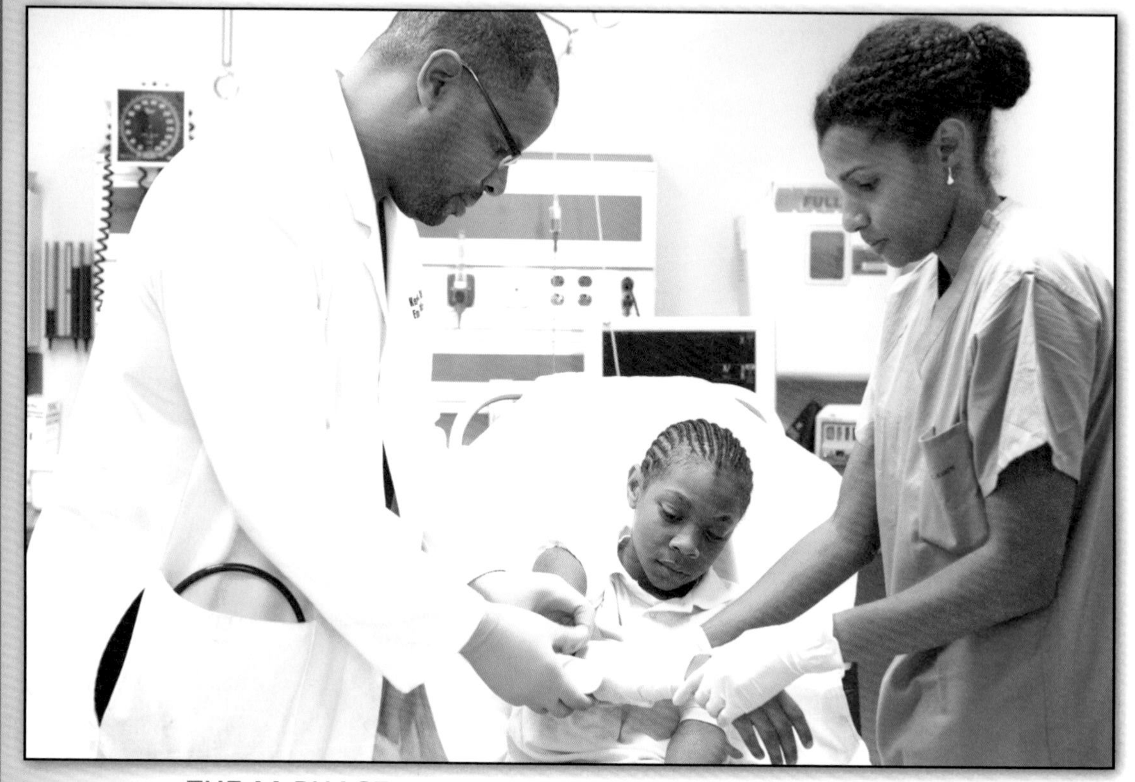

This boy has broken his arm. Healing depends on mitosis. Cells divide to fix the break. Mitosis ensures that the cells contain DNA that is identical to all other body cells.

THE M PHASE

G2 is followed by mitosis, or the M phase. Mitosis allows tissues to grow or repair by creating new cells. Mitosis involves the separation of genetic material into two new sets.

Each new cell formed through mitosis needs a complete set of the organism's DNA; it must be as similar as possible to the DNA of every other body cell.

Mitosis is itself divided into four phases. The first is called prophase. During prophase DNA in the nucleus coils up with proteins to form structures called chromatids. Pairs of sister (identical) chromatids link at their midpoints to form chromosomes.

THE G1 PHASE

Medical researchers are looking for chemicals that end the cell cycle prematurely during the G1 phase. These substances might help in the war against cancer. Cancer cells grow and divide uncontrollably. Cells might become cancerous because of an error in the G1 chemical mechanism that controls cell growth and division. It normally stops any further cell growth. More research is required, but these control substances may one day offer a cure for a range of cancers.

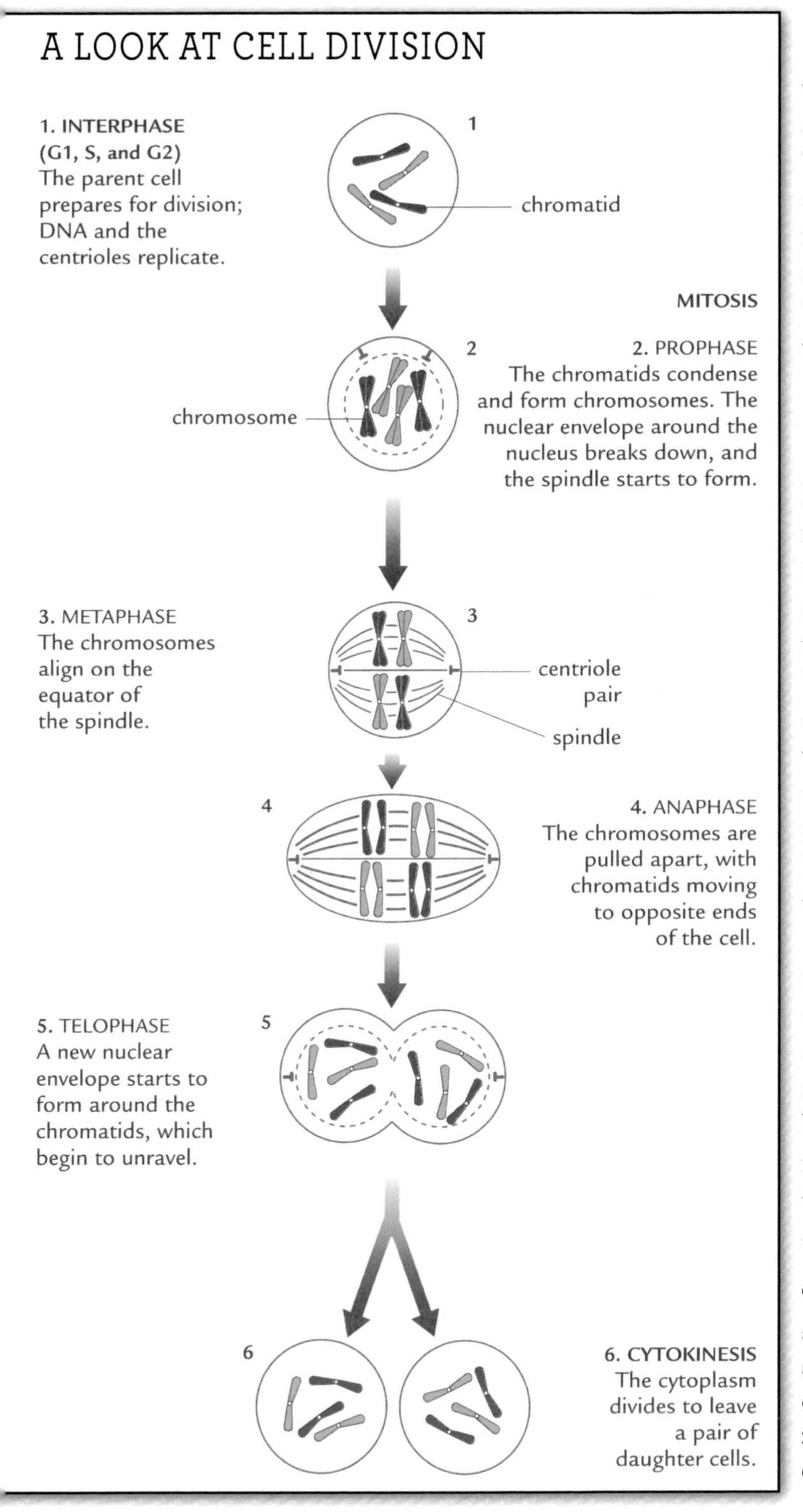

Meanwhile, tiny structures called centrioles begin to assemble outside the nucleus. The centrioles move to opposite ends of the cell, and a network of fibers called the spindle grows between them. The membrane of the nucleus then begins to break down.

METAPHASE

Next, in the phase called metaphase the chromosomes line up at the center of the cell, or equator. There they rest on the spindle. The next stage, anaphase, sees the cell's chromosomes wrenched apart by the spindle fibers. The chromatids from each of the pairs move in opposite directions toward the ends of the cell.

During this final stage of mitosis the spindle begins to disintegrate. The chromatids (or daughter chromosomes, as they are now known) begin to unravel. A new nuclear membrane begins to form around each set of chromosomes.

The stages of mitotic cell division. Note that this diagram shows division in an animal cell. Things go in a similar way in plants, but cytokinesis is different.

CELL DIVISION

Mitosis serves only to separate the two copies of the cell's DNA. Once this has been accomplished, the cell itself must physically divide. Biologists call this process cytokinesis. It involves the division of the cell cytoplasm into two. Animal and plant cytokinesis takes place in different ways.

Animal cells divide by first forming a dent along the cell equator. This is called the equatorial furrow. Tiny fibers in the cytoplasm make the cell membrane pinch inward, forming a waist in the center of the cell. The furrow quickly deepens into a groove and carries on constricting until the cell cleaves in two. Finally, a new cell membrane forms, separating the two daughter cells.

CHROMOSOMES

German biologist Walther Flemming (1843–1905) perfected observation techniques for cells. Flemming fixed cells at various stages of their life cycle. Then he stained them so he could observe the parts inside. Using dyes, Flemming could watch the chromosomes that became visible during cell division. His techniques were so efficient that he was able to see how the chromosomes divided between the daughter cells. Flemming named the division of body cells mitosis, from a Greek word meaning "thread." This referred to the threadlike appearance of chromosomes.

THE CELL CYCLE

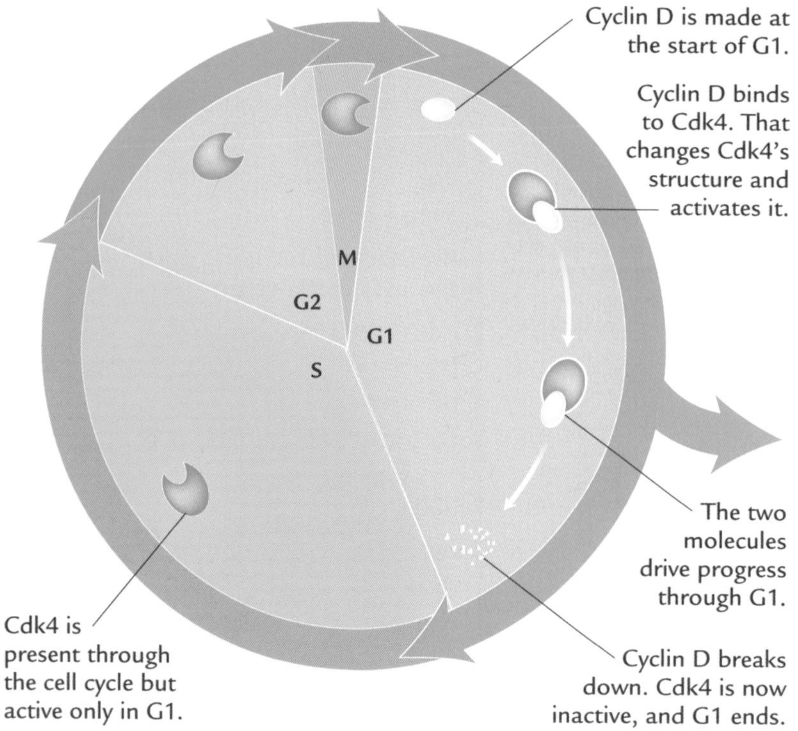

Cyclin D is made at the start of G1.

Cyclin D binds to Cdk4. That changes Cdk4's structure and activates it.

M

G2

G1

S

The two molecules drive progress through G1.

Cdk4 is present through the cell cycle but active only in G1.

Cyclin D breaks down. Cdk4 is now inactive, and G1 ends.

Different Cdk and cyclin molecules are important at different points of the cell cycle: Cdk2–cyclin E acts at the start of the S phase; Cdk2–cyclin A acts during the S phase; and Cdk1–cyclin B acts at the G2–M boundary.

The cell cycle is regulated by a protein called Cdk. It binds to another chemical called cyclin to trigger a new phase. Different types of Cdk and cyclin control different parts of the cell cycle.

UNDERSTANDING PLANT DIVISION

Plants cells have rigid walls, so they are not easily pinched inward. During telophase in plants dense material collects on a structure called a phragmoplast. It forms a double plate along the cell equator. The double plate grows until the cell is completely divided into two cells. When cell division is complete, each daughter cell has a fully formed cell wall at the site of the equator.

In both animals and plants the cell cycle begins again once cytokinesis is complete. The cell goes through the three stages of interphase until it is again time to divide.

WHAT ARE PHRAGMOPLASTS?

To successfully divide, a plant cell needs a structure called a phragmoplast. A phragmoplast is a series of fibers that forms along the midline between dividing cells. Packages (or vesicles) containing cellulose for building new cell walls are carried by proteins to the phragmoplast. They travel along the spindle fibers formed during mitosis. At the phragmoplast the packages rupture. The cellulose is then arranged by the fibers.

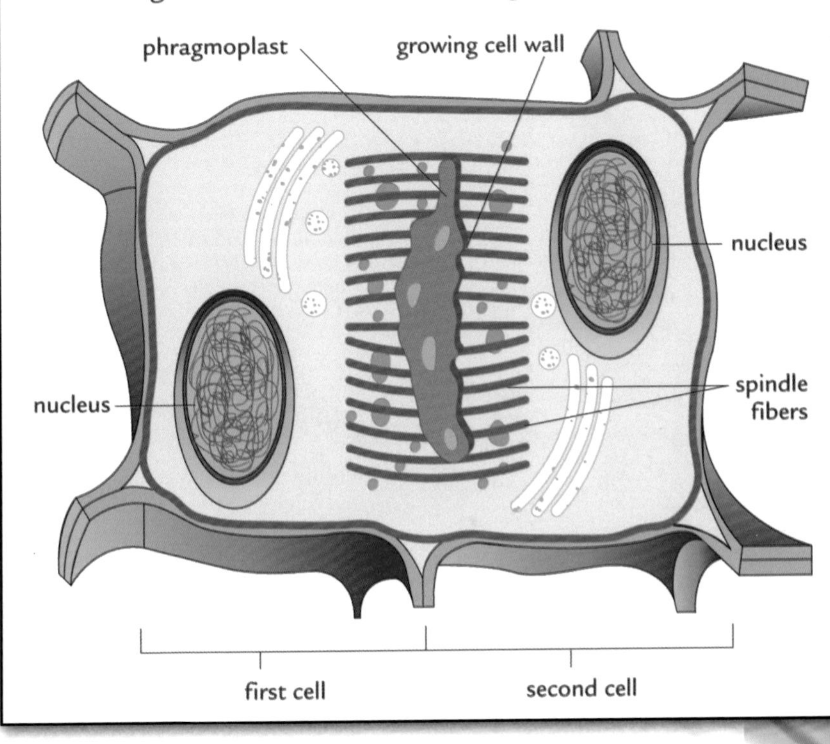

BUCKING THE TREND

Not all cells follow this typical life cycle. Nerve cells and fingernail cells buck the trend. So do red blood cells. They do not have nuclei, so they cannot divide. Red blood cells are produced by blood stem cells. They are cells in the bone marrow whose sole purpose is to create new blood cells.

Blood stem cells reproduce extremely rapidly because red blood cells must be replaced constantly. Yet the destiny of a cell produced by blood stem cells is not predetermined. Many develop into red blood cells. But other daughter cells become white blood cells, while the rest divide again and again to produce more blood stem cells.

SEX CELL DIVISION

Like blood cells, sex cells (sperm and eggs) do not go through an endless cycle of growth followed by division. In organisms that reproduce sexually, the sex cells contain half the normal number of chromosomes. Sex cells are produced by sex stem cells that occur in the testes (in males) or ovaries (in females). The process of division that leads to sex cells is called meiosis.

Meiosis is similar to mitosis in some ways, but its aim is very different. Mitosis leads to cells that are genetically identical to their parents. Meiosis produces sex cells that are genetically distinct from the parent stem cells. Sex cells cannot divide. A few unite with sex cells from mates to produce young. All the other sex cells die.

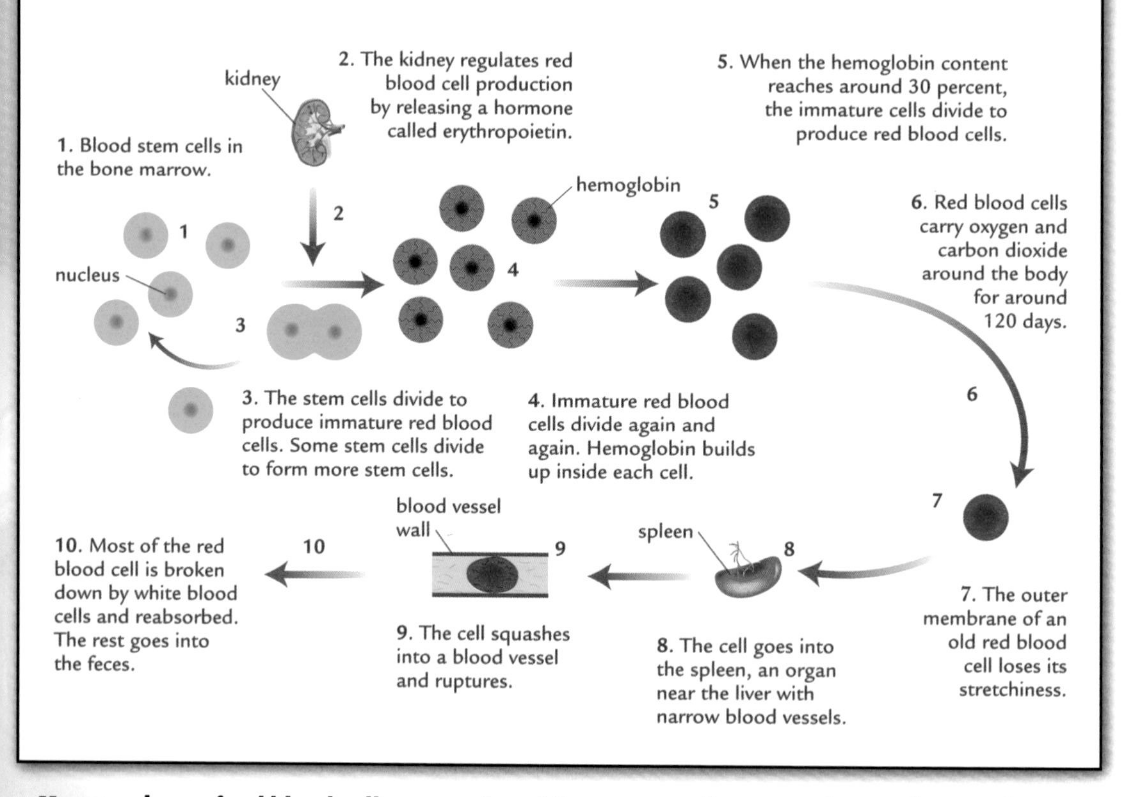

Vast numbers of red blood cells are produced from stem cells—around 2.5 million each second! Red blood cells do not have nuclei, so they cannot divide. After a few months they are broken down inside the spleen. The cells contain a pigment called hemoglobin that allows them to carry oxygen around the body.

STEM CELLS

Your body contains some cells that will stay alive without dividing for the rest of your life. Some of these cells developed when you were an embryo in your mother's uterus. Nerve cells are among these unchanging cells. When nerve cells are damaged through injury, it is extremely difficult to regrow them because such cells do not normally divide. Medical researchers are looking for ways to trick nerve cells into dividing so damaged nerves can heal. The researchers hunt for chemicals that will trigger cell division or study how stem cells work. Such research may one day allow people paralyzed by spinal cord injuries, for example, to live normal lives.

The movie actor Christopher Reeve was paralyzed from the neck down in a 1995 horse-riding accident. He died in 2004 from complications related to his injuries.

CHAPTER SEVEN

ABNORMAL REPRODUCTION

Cells sometimes reproduce and grow abnormally, resulting in diseases such as cancer.

Since you were born, your body has grown steadily and in a controlled way. You never found that your left arm had grown much longer than your right, or that your right foot had become six shoe sizes bigger than your left foot. Your arms and legs elongate at the same time and at the same rate. The genetic (inherited) information in the cells of your body ensures that your growth is properly timed and controlled.

Your body's cells also develop, reproduce, and die in a controlled way. Sometimes, however, the genetic mechanisms that regulate cell behavior fail. Then cells may grow uncontrollably, and the disease called cancer might develop. Cancer occurs when a mass, or tumor, develops from a single cell growing out of control.

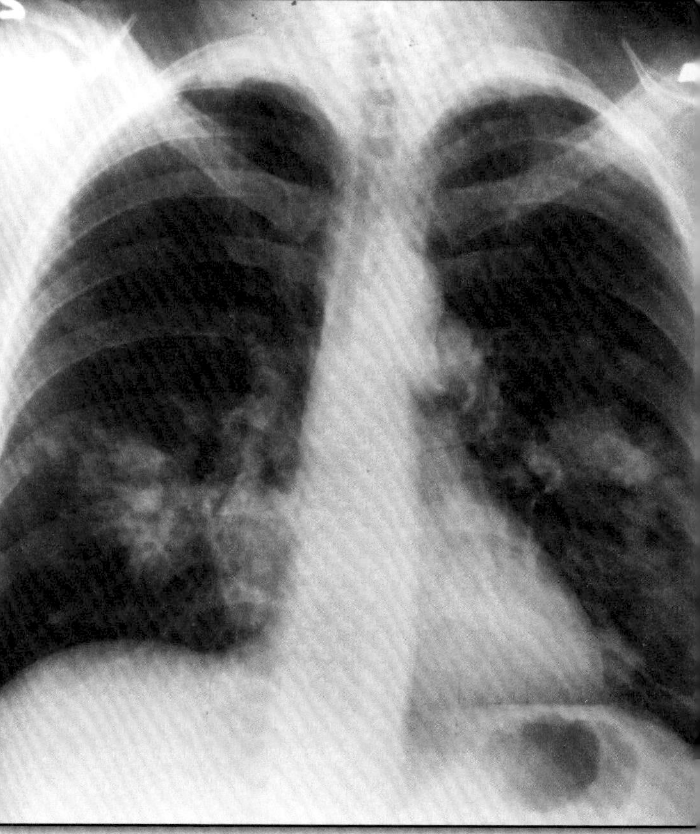

A chest X-ray reveals a possibly cancerous tumor in the patient's left lung.

GENES AND CANCER

Genes are codes that regulate the making of proteins. Many of these proteins do the work. A chest X-ray reveals so-called "cannonball" cancerous tumors in the patient's lungs. To regulate cell growth, proteins either initiate (start) or inhibit (stop) cell division. Proto-oncogenes code for proteins that switch cell division on, while tumor-suppressor genes code for proteins that switch cell division off. Cancer may develop when a mutation (change or fault) occurs in either of these types of genes.

A mutation in a proto-oncogene might cause a cell to divide endlessly, forming a tumor. A tumor-suppressor gene may be disabled by a mutation, so cell division is not ended but continues indefinitely, producing a tumor. Genes that cause cancer are called oncogenes.

Genetic mutations happen all the time. But every cell has the ability to repair damaged or mutated genes. Most of the time this repair mechanism works well.

RUDOLF VIRCHOW

In the 1850s German biologist Rudolf Virchow (1821–1902) developed his single-cell theory of tumors. He stated that all the cells in a tumor originated from a single cancerous cell. As that cell divides, it creates cancerous copies of itself. Virchow was correct, and his theory still underlies research into cancer as a disease of cells.

THE DIVISION PROCESS

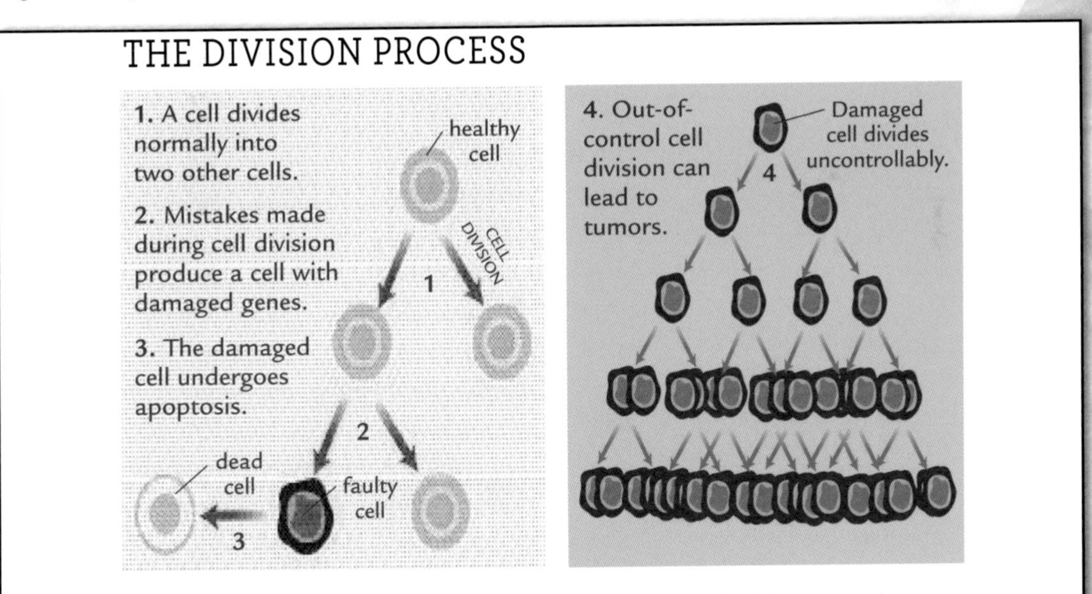

If a cell with healthy genes is damaged, the cell either repairs itself or undergoes apoptosis (cell suicide; 1–3). If the genes that control cell division fail, however, cells divide uncontrollably (4).

AN INCREASE IN CANCER

Since 1973 the incidence of many types of cancers has increased dramatically. For example, there has been about a doubling of rates of testicular cancer (from 3.3 to 5.8 per 100,000) and prostate cancer (from 80 to 175 per 100,000). Breast cancer rates have increased from 100 to 138 per 100,000; melanoma (a deadly skin cancer) from 7.6 to 26 per 100,000.

The testes and the prostate are organs of the male reproductive system. Some scientists blame the increase in reproductive cancers on artificial hormone-mimicking chemicals in the environment Yet few governments, businesses, or individuals seem to be taking action to reverse these trends by trying to eliminate possible cancer-causing agents from the environment.

Other factors can also cause the increase in rates of cancer. For example, people live longer due to improved medical care and are more likely to get cancer as they age. What do you think are the main causes?

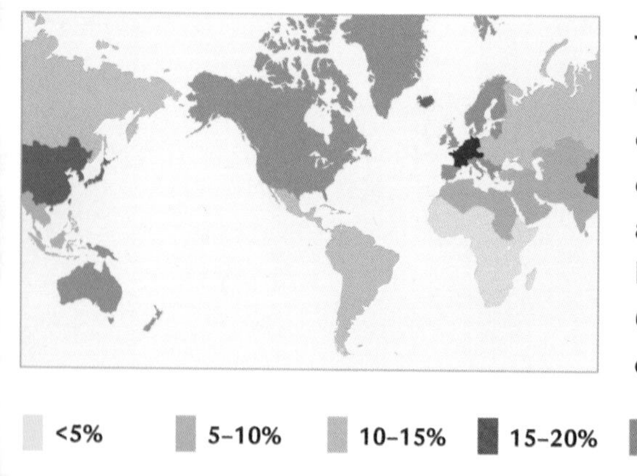

The map shows the percentage of deaths due to cancer in different areas of the world. In 2000 more than 6.2 million people died from cancer.

<5% 5–10% 10–15% 15–20% 20–25% >25%

If the genetic damage is massive, a cell can even commit suicide. Cell suicide, or apoptosis prevents damaged cells from-reproducing. It is only when cell repair or apoptosis fails that mutations may cause cell division to spin out of control.

WHAT CAUSES MUTATIONS?

Sometimes the offspring of a sexually reproducing animal inherits a mutated gene from one or both parents, and that gene may lead to cancer. A mutation in the RB gene, for example, causes an inherited form of eye cancer called retinoblastoma in children. Normally RB is bound to and controls a protein, E2F, that starts cell division. If the RB gene mutates, E2F is neither bound nor regulated, so it causes uncontrolled division of eye cells, leading to cancer.

VIRAL ONCOGENES

In 1968 researchers studying a virus discovered that when it inserts its genes into a cell, the genes in the cell are mutated, and the cell becomes cancerous. This viral gene was called src, the first oncogene identified. Human papilloma virus (HPV) might cause cancer of the cervix (the "neck" of the uterus. The virus hepatitis B might cause liver cancer. These viruses force infected cells to reproduce wildly. In the course of this

People sunbathing on a beach may not be aware of the dangers. Skin cancer can develop unless sunbathers take precautions. They should wear hats and T-shirts, avoid being in the sun during the hottest hours of the day, and use effective sunscreens to block out harmful radiation from the sun.

reproductive frenzy infected cells can undergo genetic mutations that eventually lead to cancerous tumors.

TRIGGERS

Chemicals used to make plastics, such as vinyl chloride, are known to cause mutations that cause liver cancer. Many genes help control cell division, thereby preventing cancer. However, different environmental factors can damage genes that control cell division. Damage of just a single gene rarely leads to cancer because there are other genes to serve as "backup." Damage of several key genes, by chemicals, foods, radiation, and other factors, can allow cancer to occur. Ultraviolet (UV) radiation from the sun is increasing due to destruction of the atmospheric ozone layer. This radiation

CANCER IN ANCIENT PEOPLES

Although cancer is often thought of as a modern disease, it did afflict ancient peoples. Scientists have found 5,000-year-old mummies in Egypt and Peru whose skeletons show definite signs that these individuals suffered from cancer.

causes genetic mutations in skin cells and has led to a near epidemic of skin cancer (melanoma) in many parts of the world. Exposure to arsenic, coal, and tar can also cause melanoma.

MUTATIONS

Mutations alter a gene's ability to send the signals that control cell functions. Mutations can turn normally signaling genes into oncogenes. When that happens, the normally signaling gene is said to have been "switched off" and the oncogene to be "switched on."

A normal proto-oncogene works throughout the cell cycle, controlling the timing of each stage of a cell's life. Proto-oncogenes control cell division by a cascade of signals that tell the cell when to produce the proteins needed for cell division. Similarly, tumor-suppressor genes stop cancerous cell division through a series of genetic signals. An unfixed mutation in one of these genes during any part of the cell-signaling cascade can disrupt cell division, and cancer may result.

TUMOR-SUPPRESSOR GENES

The role of tumor-suppressor genes is not to suppress tumors but to inhibit or stop cell division. Yet when functioning properly, these genes suppress tumor growth by stopping cell division.

One of the best-studied tumor-suppressor genes is called p53. p53 occurs in the cell's nucleus, where it monitors deoxyribonucleic acid damage. Segments of deoxyribonucleic acid, or DNA, form

Cancer is generally a multistep process in which a series of mutations (changes) results in the production of cancerous cells. Key stages are those that affect the ability to regulate cell division.

SWITCHING ERRORS

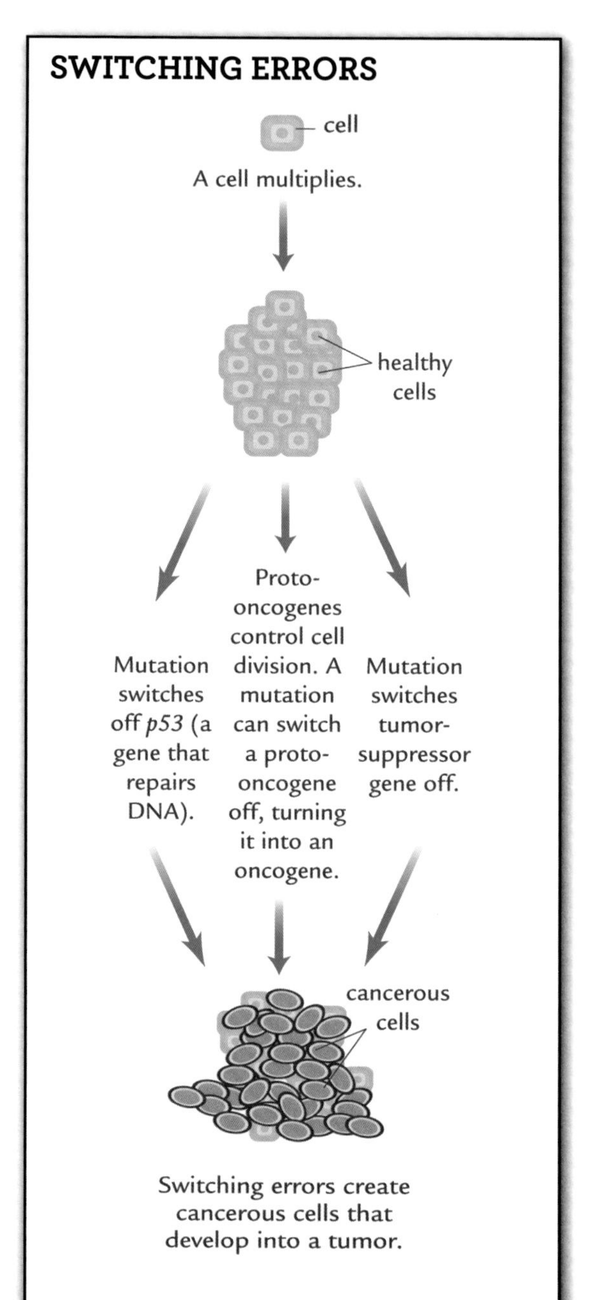

Switching errors create cancerous cells that develop into a tumor.

DR. GEORGE PAPANICOLAOU

Early detection of cancer can save lives. A Pap smear is a simple, painless test named after its developer, Dr. George Papanicolaou (1883–1962). Pap smears can detect cervical cancer during its earliest stages. Because many women have Pap smears every year or two, deaths from cervical cancer have decreased in recent years.

genes. Slight DNA damage causes p53 to trigger the cell's repair mechanisms. If there is severe damage, p53 initiates apoptosis, or cell suicide. The p53 gene may itself suffer mutations that switch the gene off. Cells with a damaged p53 gene cannot repair their DNA nor self-destruct. p53 mutations are present in 70 percent of colon cancers and up to 50 percent of breast and lung cancers.

p53 is not the only gene that controls apoptosis. A mutation in the proto-oncogene bcl-2 causes it to produce excessive quantities of a protein that blocks apoptosis. Another mutation turns on this protein permanently, so cell suicide becomes impossible.

HOW CANCER TRAVELS

A cancer that grows in a vital organ can be fatal if it destroys that organ. Cancer can be fatal even if it originates in a non-vital organ because cancer can spread, or metastasize, through the body. To spread, the cells in a cancerous, or malignant, tumor must accomplish some formidable tasks. First, they must break away from other cells; second, they must break through surrounding membranes and tissues; and third, they must find, enter, and travel through blood vessels.

All cells are embedded in the jellylike extracellular matrix. This matrix contains cell adhesion molecules, or CAMs, which hold cells together. When a cancer cell loses its CAMs, it is freed from other tumor cells and from the extracellular matrix. Then the cancer cell produces

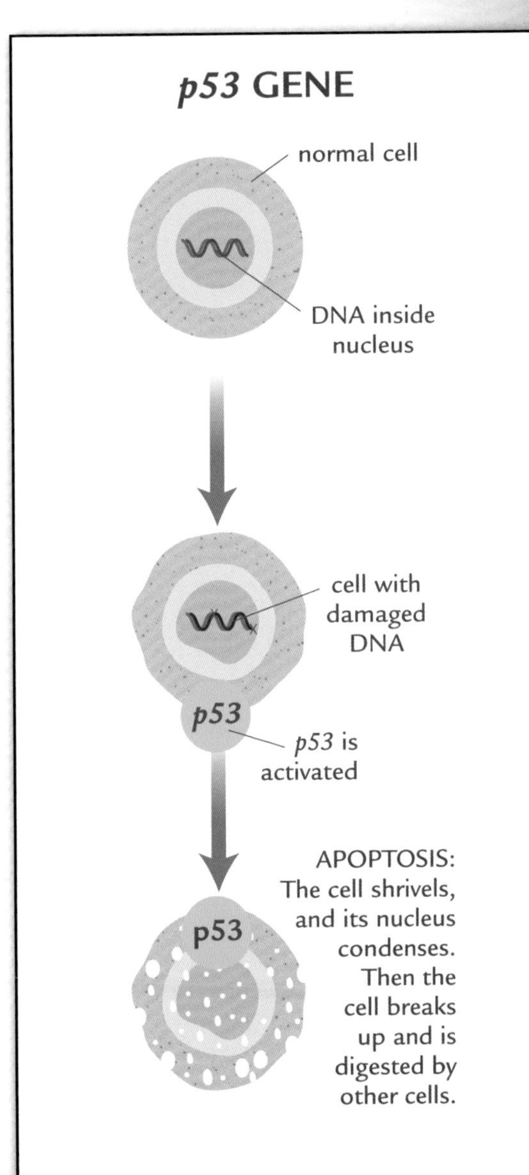

ASPARTAME

Since 1973 the incidence of brain cancer has been increasing at an overall rate of about 0.7 percent annually. Researchers have linked this increase to the artificial sweetener aspartame, which is widely used in diet foods and beverages. The researchers have been pressuring the U.S. Food and Drug Administration (FDA) to ban the use of aspartame, but so far the agency has refused. Some researchers believe that the FDA's refusal arises from its close ties to the food industry. There may be other causes of increasing brain cancer, such as radiation from electronic devices.

enzymes that help it break through the membranes and tissues that block its path to the bloodstream.

Cancer cells invade tissues and organs by migrating into them directly. Cancers spread when cancer cells enter the bloodstream and are carried to distant organs, in which they then reproduce. The first tumor is called the primary cancer; if cancerous cells metastasize to elsewhere, those tumors are called secondary cancers.

BLOOD VESSEL FORMATION

Angiogenesis is the formation of blood vessels in a tumor. It is the stage when a

HOW CANCER MOVES THROUGH THE BLOODSTREAM

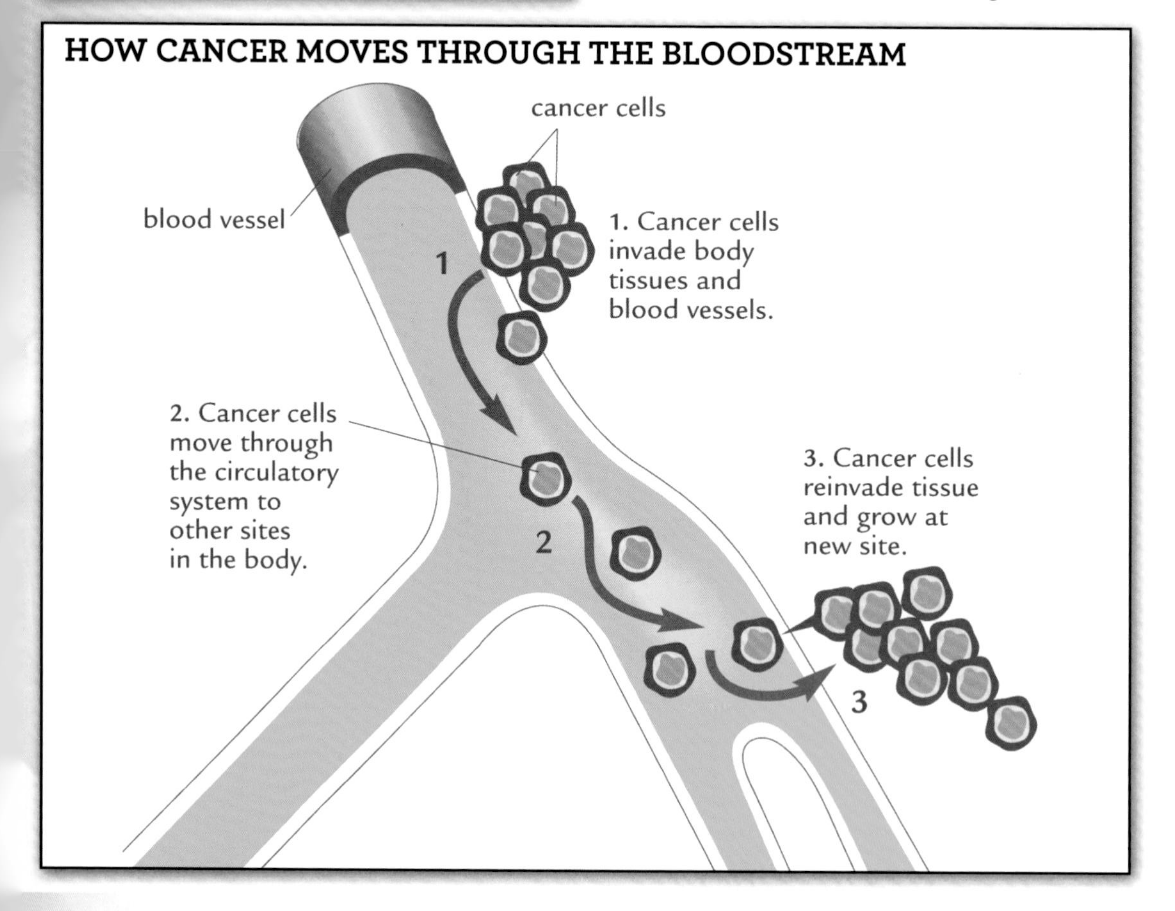

tumor turns from a mass of cells into a malignant growth. Cancer cells make proteins that help build blood vessels, in particular, vascular endothelial growth factor, or VEGF. This protein attracts endothelial (lining) cells into the tumor. Endothelial cells build blood vessels. Inside the tumor they create capillaries, which bring vital nutrients and oxygen to the tumor. So VEGF enables cancer cells to build their own highway out of the tumor and into the circulatory system.

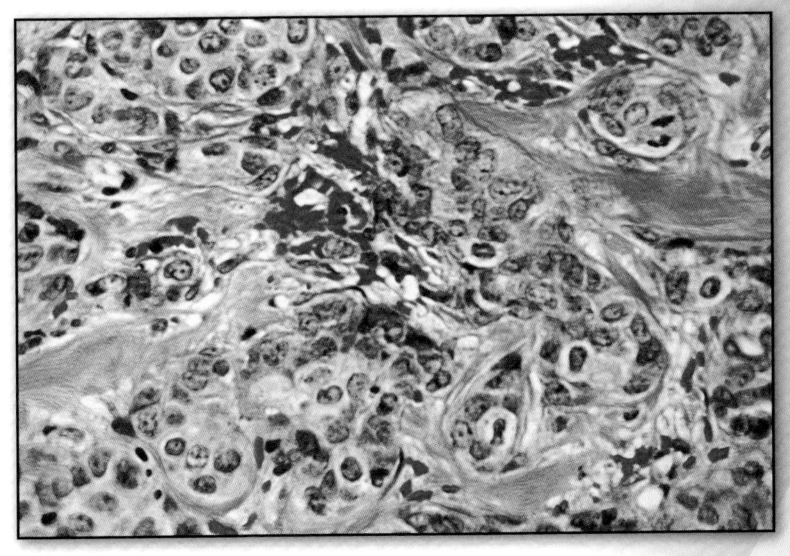

The dark purple cells in this image are breast cancer cells. Breast cancer is one of the most common cancers affecting women. Prolonged exposure to the hormone estrogen is a likely cause. This can result from an early puberty or a late menopause. Obesity may also be responsible, since excess fat causes extra estrogen to be produced. Treatment for breast cancer may involve surgery, chemotherapy, and radiotherapy.

THE MYSTERY OF METASTASIS

Scientists do not fully understand how and why different primary cancers metastasize to some organs and not others. Some cancer cells are carried around the body by the blood. After the blood has left most organs, it is pumped to the lungs by the heart. So the lungs are often the first site of metastatic cancer.

Researchers have found that some cancer cells have an attraction for cells in other parts of the body. For example, cells from breast and prostate cancer often metastasize first to bones. Scientists think that in these organs there is an attraction between receptor proteins on the surface of cancer cells and molecules in the extracellular matrix, which fills spaces among cells.

UNDERSTANDING GENE THERAPY

Gene therapy addresses the genetic mutations that lead to cancer. Scientists plan to use recombinant DNA technology to replace a defective, mutated gene with its normal, healthy counterpart. The healthy gene is then inserted into cancer cells to restore the normal control of their reproduction.

THE VARIOUS CANCER TREATMENTS

Surgery is used together with treatments such as radiotherapy and chemotherapy.

TREATMENT THROUGH SURGERY

Cutting a cancerous tumor out of the body is the most direct way of ridding the body of a malignant tumor. Until recently surgery was the only method of cancer treatment.

Surgery can be problematic because it is never certain that all the malignant cells have been removed. Physicians estimate that in two-thirds of the patients undergoing cancer surgery the cancer has already spread beyond the surgical site.

THE RADIOTHERAPY ROUTE

Bombarding cancer cells with radiation destroys them in one of two ways: Either the cells are so damaged they cannot reproduce, or the radiation damage causes apoptosis.

There are two types of radiation treatment. In internal radiotherapy a radioactive material is inserted into the body near the tumor. External radiotherapy is when a beam of radiation is directed at the site of the tumor from an apparatus outside the body. The radiation is administered in tiny pulses that minimize complications and side effects.

KILLING CANCER CELLS THROUGH CHEMOTHERAPY

Chemotherapy uses drugs to kill cancer cells. Most of these chemicals prevent cancer cells from multiplying. One advantage of chemotherapy is that the chemicals travel throughout the body and can kill cancer cells almost anywhere they are found. Most chemotherapy patients suffer side effects such as vomiting and hair loss because the chemicals often kill healthy cells as well as cancerous ones.

INNOVATIVE TREATMENTS

Today, physicians use their greater understanding of how cancer cells function to fight the disease. One new treatment is called angiogenesis inhibition. It involves giving the patient drugs to prevent angiogenesis. Immunotherapy uses the body's own immune system to fight the cancer. The immune system's B-cells recognize particles called antigens on the surface of foreign cells and build proteins

PREVENTION

Cancer is not inevitable. People can avoid behaviors that are known to make them susceptible to cancer. For example, you can choose not to use tobacco, not to eat much red and fatty meat, and not to use household or garden products containing known cancer-causing agents. To reduce the risk of getting skin cancer, avoid going out in strong sun. When outdoors, always wear a hat and light clothing, and use an effective sunscreen.

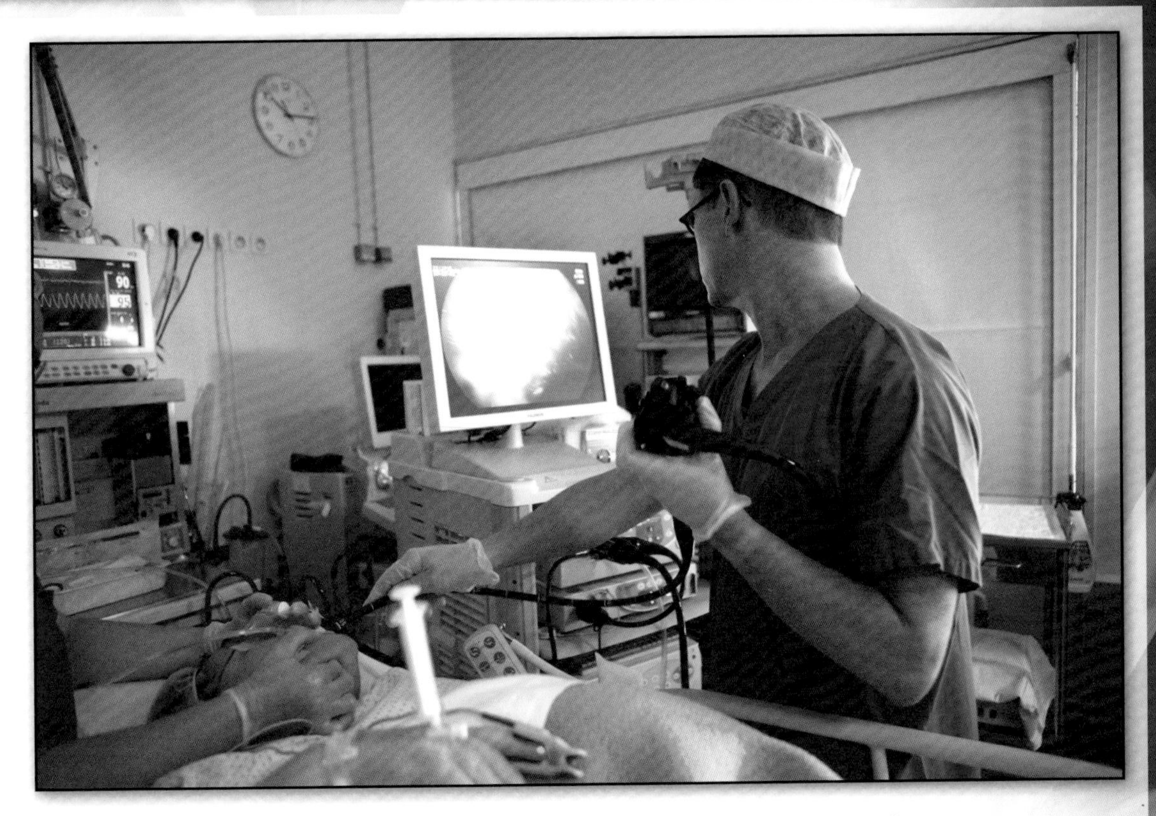

Physicians perform an endoscopy. An endoscope is a fiber-optic cable with a tiny camera at one end. It produces an image on a screen. The physicians remove a sample of tissue (biopsy) from the stomach and use it to diagnose cancer.

called antibodies to destroy the cells. (Antigens are anything that provoke an immune system response.) Researchers have also identified antigens on the surfaces of cancer cells. Now scientists are finding ways to create antibodies that kill tumor cells. Another recent approach involves using vaccines to stimulate the body's own immune system to attack and kill cancer cells.

CHAPTER EIGHT

BIOGRAPHIES: FRANCIS CRICK, ROSALIND FRANKLIN, AND JAMES WATSON

It was the work of English X-ray crystallographer Rosalind Franklin, and of the team of English molecular biologist Francis Crick and American biologist James D. Watson, that led to the discovery of DNA's structure and how it copies itself.

Rosalind Franklin was born in London on July 25, 1920. Her parents sent her to one of the few girls' schools in London that taught physics and chemistry. By the time she was 15 years old Rosalind had decided to become a scientist. Although her father was initially opposed to the idea, he eventually agreed, and in 1938 she entered Newnham College, Cambridge, to study chemistry and physics, graduating in 1941. After graduating, she stayed on at Cambridge for a year to do postgraduate research, but it was wartime and in 1942 she joined the British Coal Utilization Research Association to do war work. In 1945 Cambridge University awarded her a Ph.D. in physical chemistry.

Rosalind Franklin was instrumental in the work of Francis Crick and James Watson and their research into the structure of DNA.

STUDYING THE STRUCTURE OF CRYSTALS

In 1947 Franklin moved to Paris and spent the following three years working at the Central Government Laboratory for Chemistry. These were happy and productive years, during which she learned X-ray diffraction (also known

The aerial photograph shows some of the colleges that make up Cambridge University, England. King's College, with its chapel, is at front. Franklin graduated from Newnham College, Cambridge; Crick was a research student at Caius College; and Crick and Watson carried out their work on the structure of DNA at the university's Cavendish Laboratories.

KEY DATES	
1916	Francis Crick is born on June 8
1920	Rosalind Franklin is born on July 25
1928	James Watson is born on April 6
1937	Crick graduates in physics from University College, London
1941	Franklin graduates in chemistry and physics from Newnham College, Cambridge
1943	Watson enrolls at Chicago University
1945	Cambridge University awards Franklin a Ph.D. in physical chemistry
1947	Crick joins the Strangeways Research Laboratory in Cambridge; Watson gets degree in zoology and moves to Indiana University
1947–50	Franklin works at the Central Government Laboratory for Chemistry in Paris
1949	Crick joins the Medical Research Council unit at the Cavendish Laboratory, Cambridge
1950	Crick becomes a research student at Caius College, Cambridge; Watson is awarded his Ph.D.
1951	Watson goes to work at the Cavendish Laboratory and meets Crick; Watson learns X-ray diffraction techniques and works with Crick on the problem of DNA structure
1951–53	Franklin joins a research group at King's College, London under Maurice Wilkins, working on X-ray diffraction of DNA
1953	Wilkins shows Franklin's DNA photographs to Crick and Watson, who publish their discovery of the structure of DNA on April 25
1955	Watson joins biology department at Harvard University, becoming professor in 1961
1958	Franklin dies on April 16
1962	Crick, Watson, and Wilkins receive the Nobel Prize for physiology or medicine
1968	Watson becomes director of the Cold Spring Harbor Laboratory, Long Island, N.Y.
1977	Crick becomes Kieckhofer Professor at the Salk Institute for Biological Studies in San Diego, California
1988	Watson is appointed associate director for human genome research for the National Institutes of Health
1989	Watson is made director of the National Center for Human Genome Research and launches the Human Genome Project
1994	Watson becomes president of the Cold Spring Harbor Laboratory

as X-ray crystallography) techniques. These would prove vital in the search for the structure of the DNA molecule. Molecules, even very big ones, are far too small for a chemist to be able to examine them and analyze their structure under the microscope. Chemists can work out from the rules of chemistry how each atom is attached to its neighbors. What they cannot work out is the final shape of a molecule, which can contain hundreds of atoms. X-ray crystallography can reveal a molecule's shape. A crystal is an arrangement of regularly spaced atoms. When a beam of X-rays passes through a crystal, each of the atoms in the crystal deflects, or bends, the beam in a characteristic way caused by the density of electrons in each atom. Electrons are particles with negative charge that surround each atom's nucleus.

X-rays will expose photographic film, so if a piece of film is placed behind the crystal, the deflected X-rays will form a pattern of spots that will appear when the film is developed. The pattern of spots reveals the density of the electrons in each part of the picture. Knowing the electron density makes it possible for scientists to calculate the position of each atom in the crystal. This can then be plotted as a computer graphic, using three coordinates to specify each position, or used to build a model using colored balls and wires.

In 1951 Franklin returned to England to become a research associate in the biophysics research laboratory at King's College, London. It was there that she met Maurice Wilkins, who was then assistant director of the laboratory. Franklin and Wilkins were both set to work on examining the DNA molecule, but Wilkins assumed authority over her, and this soured the relationship between them.

MAURICE HUGH FREDERICK WILKINS 1916–2004

Maurice Wilkins was born on December 15, 1916, in Pongaroa, New Zealand. His father was a physician who was originally from Ireland. When he was six the family moved to England and Maurice was educated at King Edward VI School, Birmingham, and St. John's College, Cambridge. He graduated with a degree in physics in 1938 and obtained his Ph.D. from the University of Birmingham in 1940. He spent the remainder of the war working first on radar and then, at the University of California, on the Manhattan Project to develop the atomic bomb. His interest then turned to biophysics and in 1945 he obtained a post working on biophysics at St. Andrew's University, in Scotland. In 1946 he joined the Biophysics Research Unit of the Medical Research Council at King's College, London; it was here that he began to study the structure of the DNA molecule with the help of X-ray diffraction photographs taken by his colleague Rosalind Franklin. For his work in helping Crick and Watson unravel the structure of DNA, Maurice Wilkins shared in the 1962 Nobel Prize for physiology or medicine. Anxious that advances in science should be closely monitored, he has served as president of the British Society for Social Responsibility in Science (1969), and of the Russell Committee against Chemical Weapons.

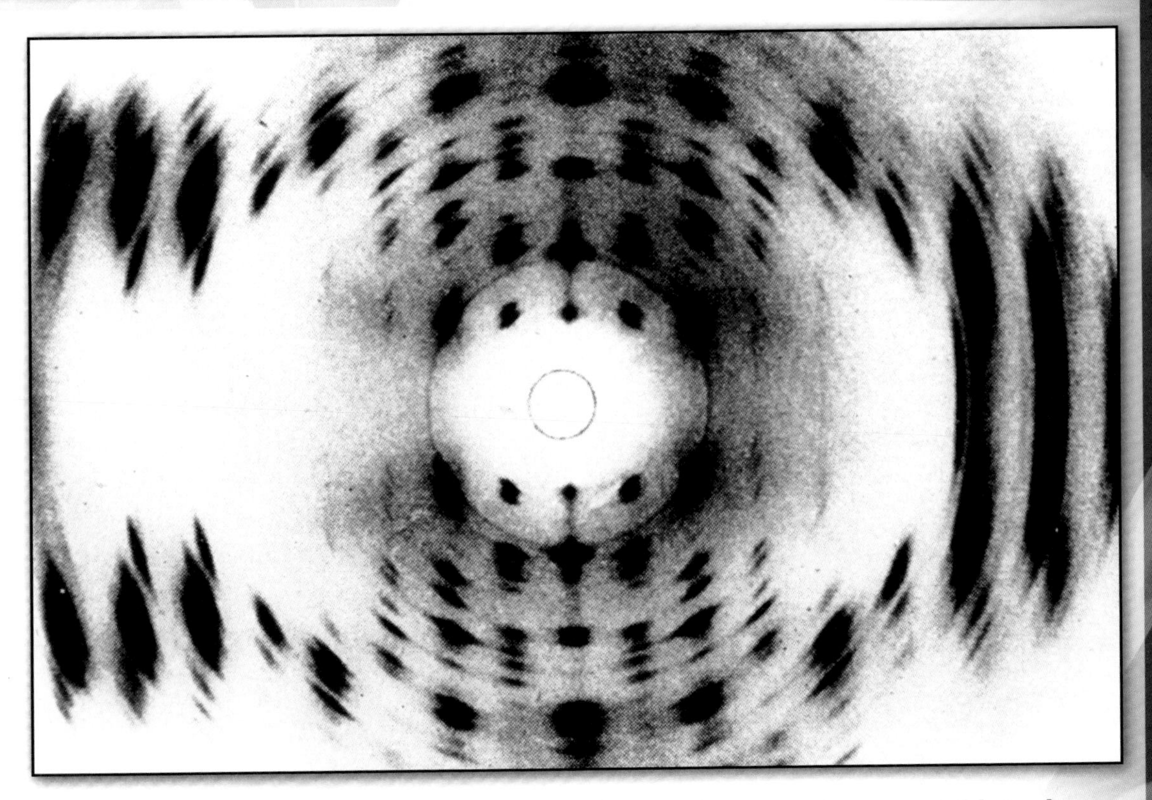

An X-ray diffraction image of part of a crystal of DNA. The repeated pattern visible in photographs such as these played a major role in identifying the structure of DNA.

UNDERSTANDING DNA

New individuals resemble their parents. If you plant an acorn it will grow into an oak tree and never an elm tree. Cats give birth to kittens, and never to puppies. In 1865 biologist Gregor Mendel (1822–1884) presented a paper in which he described experiments he had conducted with pea plants. Mendel had discovered the way particular characteristics are transmitted from parents to their offspring by means of what he called "heritable factors."

In 1882 the German biologist Walther Flemming (1843–1905) described seeing these "heritable factors" as short, threadlike objects that he called "chromosomes." Flemming had no idea of the importance of chromosomes because he had not heard about Mendel's work. Ten years later the German zoologist August Friedrich Leopold Weismann (1834–1914) proposed that every living organism contains "germ plasm," a special substance that controls the development of every part of the body and that is passed from parents to their offspring. Weismann also realized that if germ plasm from both parents is mixed at fertilization, the amount of it must increase from one generation to the next. Since this does not happen, there must be a type of cell division in which the resulting cells receive only half the full amount of germ plasm. This means there must be some way in which the germ plasm can be halved.

This was confirmed by another German biologist, Wilhelm August Oskar Hertwig (1849–1922). In 1875 Hertwig had described in detail the fertilization of a seaurchin egg. He then described the type of cell division proposed by Weismann in which the amount of germ plasm is halved. This reduction-division process is now called meiosis.

In 1946 it was found that "germ plasm" is made from a nucleic acid called deoxyribonucleic acid, or DNA for short. We now know that genes (DNA) direct the cell to assemble proteins. These proteins give us our individual features.

DNA is a very complex compound and very difficult to work with. Franklin brought her skill in X-ray diffraction to bear on the problems involved, designing a camera that would provide images of DNA that had been stretched into a thin fiber. Using this technique she was the first to report that the sugar-phosphate backbone of DNA lies on the outside of the DNA molecule. Her photographs showed that DNA might be arranged as a helix—a spiral shape, like the thread of a screw.

THE WATSON AND CRICK PARTNERSHIP

Franklin was close to discovering the structure of DNA by herself, but there were two other people who would prove vital to the outcome of the story.

Francis Harry Crick was born on June 8, 1916, in Northampton, central England. In 1937 he graduated with a degree in physics from University College, London. The outbreak of World War II in 1939 prevented him from completing his doctorate at that time; instead he went to work for the British Admiralty, studying naval mines and helping to develop radar. In 1949 he joined the Medical Research Council unit at the Cavendish Laboratory, Cambridge, where he worked on X-ray diffraction of proteins. He became a research student at Caius College, Cambridge, in 1950, and was awarded his Ph.D. in 1953.

In collaboration with Rosalind Franklin and James Watson, Francis Crick helped pave the way for our current understanding of the structure of DNA.

James Dewey Watson was born on April 6, 1928, in Chicago, Illinois. He was educated in Chicago, and in 1943, at the age of 15, he won a scholarship to study zoology at Chicago University. He graduated in 1947. Then, still only 19 years old, he won a fellowship for graduate study at Indiana University, in Bloomington, where he was awarded his Ph.D. in zoology in 1950.

As a boy Watson had been an avid birdwatcher. His interest in birds, and their huge diversity in shape and color, led him to study genetics. Several distinguished geneticists were working in the bacteriology department at Indiana University, and they guided him in his doctoral research into the effect of X-rays on the multiplication of bacteriophages (viruses that attack bacteria). Watson became convinced that the chemical structure of genes was of fundamental biological importance.

Along with Francis Crick, James Dewey Watson co-discovered the structure of DNA.

Watson decided to begin investigating the structural chemistry of nucleic acids and proteins, and learned that scientists were using photographs of X-ray diffraction patterns of protein crystals to study the structure of protein molecules. He worked at the University of Copenhagen, before in 1951 moving to the Cavendish Laboratory in Cambridge, where he began working with Francis Crick on the problem of the structure of DNA. The two men complemented each other: Crick had experience with X-ray diffraction, and Watson understood genetics and biochemistry. It was at this time that Maurice Wilkins (without Franklin's knowledge) showed Crick and Watson the photographs of the X-ray diffraction patterns made by DNA crystals that Franklin had taken.

THE DOUBLE HELIX

Crick and Watson seized on the key role of DNA as the basic material of heredity; they became convinced that if the structure of DNA were known, then its role in heredity would become clear. Helped by the X-ray diffraction studies passed to them by Wilkins, Crick and Watson made

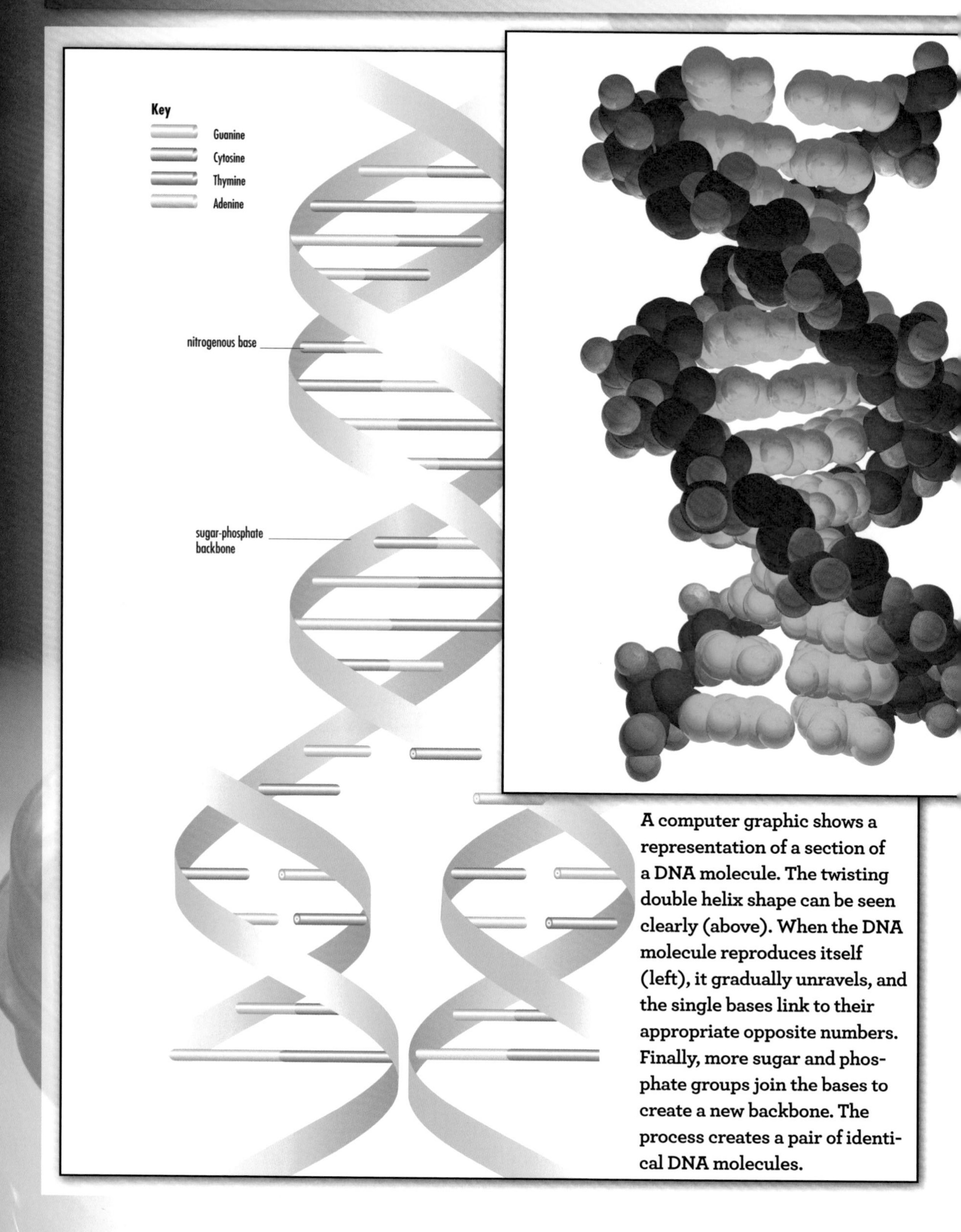

A computer graphic shows a representation of a section of a DNA molecule. The twisting double helix shape can be seen clearly (above). When the DNA molecule reproduces itself (left), it gradually unravels, and the single bases link to their appropriate opposite numbers. Finally, more sugar and phosphate groups join the bases to create a new backbone. The process creates a pair of identical DNA molecules.

up a model of DNA based on its known properties. In a moment of inspiration they realized that the DNA molecule must be arranged in a double spiral, like two long twisting ladders wound around each other—a "double helix" construction. Eventually Crick and Watson were able to show how a molecule consisting of two helixes running in opposite directions could separate into single strands and how each strand could then assemble a complementary strand. They described their findings in two papers published in the British scientific journal *Nature* in April and May, 1953.

AFTER DNA

Rosalind Franklin later began working at Birkbeck College, London, studying viruses. She worked on the tobacco mosaic virus, which causes discoloring of leaves in tobacco and other plants, and also investigated the polio virus. Tragically, in 1956 she was diagnosed with cancer, and she died two years later at age 37. Nobel Prizes are never awarded after death, so she did not share in the Nobel Prize for physiology or medicine that Crick, Watson, and Wilkins received in 1962 for their work on DNA.

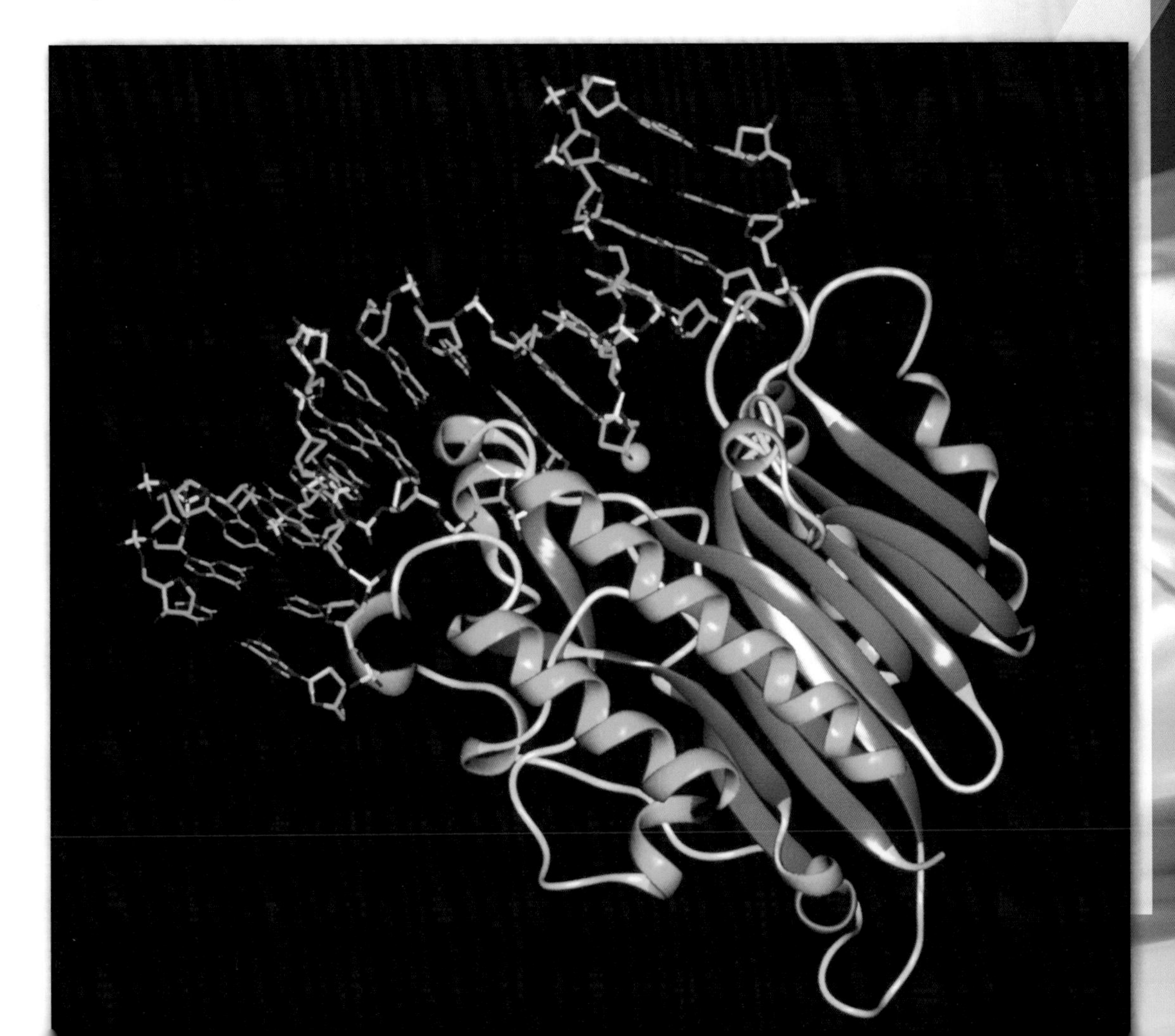

After his work on the structure of the DNA molecule, Crick did further important work on the "genetic code," the sequence of bases in DNA that provide the instructions for the production of proteins, and helped identify the groups of three bases (codons) that form the approximately 20 amino acids that make up proteins. He was a visiting professor at Harvard University in 1960. In 1977 he became professor at the Salk Institute for Biological Studies in San Diego, California. He then turned his attention to the brain; this has been his main research interest since the late 1970s.

Meanwhile, Watson returned to the United States in 1953 to take up a post as a senior research fellow in biology at the California Institute of Technology. At Caltech he made use of X-ray diffraction techniques to study RNA. In 1955 he returned to the Cavendish Laboratory, where he worked for a year with Crick, this time investigating the structure of viruses. In 1956 Watson joined the Biology Department at Harvard; he became professor there in 1961. In 1968 he became director of the Cold Spring Harbor Laboratory on Long Island, New York. In 1989 he was appointed director of the National Center for Human Genome Research; for a time he headed the Human Genome Project. In 1994 he became president of Cold Spring Harbor Laboratory. Under his administration the laboratory became one of the world's leading centers for molecular biology (the study of cell molecules), and one of the wealthiest. There Watson guided research that led to the present understanding of the molecular basis of cancer.

SCIENTIFIC BACKGROUND

Before 1940

Austrian biologist Gregor Mendel (1822–1884) presents a paper describing "heritable factors"

German biologist Walther Flemming (1843–1905) reveals chromosomes

German zoologist August Weismann (1834–1914) reports germ plasm

Wilhelm Hertwig (1849–1922), a German biologist, first describes reduction cell division (meiosis)

X-ray crystallography is developed by father and son British physicists W. H. Bragg (1862–1942) and W. L. Bragg (1890–1971)

1944 American bacteriologist Oswald Avery (1877–1955) shows that DNA isolated from one bacterium can alter the appearance of a second strain

1946 British chemist Dorothy Hodgkin (1910–1994) unravels the complex structure of the antibiotic drug penicillin using X-ray crystallography

1946 New Zealander Maurice Wilkins (1916–) joins the Biophysics Research Unit of the Medical Research Council at King's College, London, to work on the structure of the DNA molecule

1947 Franklin works at the Central Government Laboratory for Chemistry in Paris, France, on X-ray diffraction techniques

1949 Crick joins the Medical Research Council at the Cavendish Laboratory, Cambridge, to work on X-ray diffraction

1951 Watson begins to investigate the structural chemistry of nucleic acids and proteins and teams up with Crick

1953 Wilkins shows Franklin's DNA photographs to Watson and Crick, who publish their discovery of the structure of DNA on April 25

1955 Watson becomes assistant professor at Harvard University

1958 Franklin dies of cancer

1959 Crick is a visiting professor at Harvard University

1959 British virologist Hugh Cairns (1922–) succeeds in carrying out the genetic mapping of an animal virus for the first time

1962 Crick, Watson, and Wilkins receive the Nobel Prize for physiology or medicine

1968 Watson publishes his account of the discovery of the structure of DNA, *The Double Helix*

1968 Watson becomes director of Cold Spring Harbor laboratory Long Island, N.Y.

1969 Researchers identify the beginning of a single gene in strands of DNA

1973 As a result of work done by American biologists Daniel Nathans (1928–) and Hamilton Smith (1931–) on restriction enzymes, scientists succeed in implanting a gene into a bacterium

After 1975

1980 American molecular biologists Paul Berg (1926–), Walter Gilbert (1932–), and British biochemist Frederick Sanger (1918–) share the Nobel Prize for chemistry for their work on nucleic acids

1990 The Human Genome Project is set up to construct a map of human genetic composition

1990 Gene therapy on a human being is carried out for the first time

2001 A working draft of the human genome is published

POLITICAL AND CULTURAL BACKGROUND

1940 Tenzin Gyatso, (1935–) is enthroned as the 14th Dalai Lama, the spiritual leader of Tibet

1943 As World War II (1939–45) continues, in the Pacific the U.S. navy begins a campaign of "island-hopping" toward Japan—using one island after another as a base for capturing the next

1945 In Italy, fascist leader Benito Mussolini (1883–1945) is captured, tried, and executed by partisans while trying to escape to Switzerland

1947 The U.S. State Department begins the Voice of America radio station, broadcasting to the Soviet Union in Russian

1950 On July 25 troops from North Korea suddenly invade South Korea, starting a war that will involve troops from 15 United Nations (U.N.) member countries; an armistice is declared on July 27, 1953 but no formal peace treaty is ever signed

1950 The musical *Guys and Dolls* opens in New York. With songs by Frank Loesser (1910–1969), it will become one of Broadway's best-loved shows

1958 In France the proposals by General Charles de Gaulle (1890–1970) that the president should have wider powers are approved by referendum; de Gaulle becomes the first president of the Fifth Republic

1960 Democratic senator John F. Kennedy (1917–1963) defeats Republican vice-president Richard Nixon (1913–) to become the youngest person to be elected U.S. president

1963 Tamla Motown releases the first album by blind singer Steveland Judkins (1950–) under his stage name; it is called *Little Stevie Wonder: the 12-year-old Genius*

1964 American boxer Cassius Clay (1942–) wins the world heavyweight boxing championship; he announces his conversion to Islam and adopts the name Muhammad Ali

1967 In China the People's Liberation Army moves to restore order and the authority of the Communist Party as Red Guards rampage through the country

1967 South African surgeon Christiaan Barnard (1922–) performs the first human heart transplant

1971 The 26th Amendment to the U.S. Constitution gives 18-year-olds the right to vote

1972 Soviet gymnast Olga Korbut (1956–) wins three gold medals and one silver and becomes the darling of the Olympic Games held in Munich, Germany

angiogenesis The formation of blood vessels by cancer tumors.

antigen Molecule (often on the surface of a foreign body) that the immune system can recognize.

apoptosis Cell suicide triggered by tumor-suppressor genes.

autocrine signaling When a cell sends out chemical signals but receives and acts on them itself.

axon Long extension of a neuron that carries electrical signals.

cell adhesion molecule (CAM) Molecule that allows a cell to stick only to similar cells.

cell cycle Life cycle of a cell.

centriole Structure in the cell that helps form the spindle.

chloroplast Organelle that carries out photosynthesis.

chromosome Structure that forms during cell division and is made of DNA.

cilium Small filament that occurs in banks on the surface of many cells.

cyclic AMP (cAMP) Messenger chemical that drives a cell's response to a hormone.

cytokinesis The physical separation of one cell into two after meiosis or mitosis.

cytologist Biologist who studies cells.

cytoplasm Region of a cell outside the nucleus.

cytoskeleton Fibers in the cytoplasm that provide a cell with structural support.

cytosol Liquid that forms much of the cytoplasm.

desmosome Tough joint that holds cells together.

deoxyribonucleic acid (DNA) Molecule that contains the genetic code for all cellular (nonvirus) organisms.

dynein Protein that allows cilia and flagella to move.

egg Female sex cell.

endocrine signaling Long-distance signaling using hormones.

endocytosis When a cell takes in molecules or particles.

enzyme Protein that speeds up chemical reactions inside an organism.

eukaryote cell Cell of a plant, animal, fungus, or protist; contains structures called organelles.

exocytosis The process cells use to get rid of substances.

flagellum Long, tail-like structure used for locomotion by many single-celled organisms.

gap junction Tiny channel connecting two cells through which small molecules can pass.

genome All the genes present inside an organism.

Golgi apparatus Organelle that alters and packages molecules made in other parts of the cell.

Haversian canal Channel through which blood vessels bring nutrients to bone cells.

hormone Chemical messenger that regulates life processes inside the body.

hydrophilic Molecule that repels water.

hydrophobic Molecule that attracts water.

interphase Nondividing part of a cell's life cycle.

lysosome Organelle that breaks down large molecules, such as proteins.

meiosis Process of cell division that leads to sex cell production.
metastasis The movement of cancer cells from one part of the body to another.
mitochondrion Organelle that produces energy from food and oxygen.
mitosis Process of cell division that leads to the production of body cells.
mutation A change in a cell's DNA.
myelin sheath Fatty coating of a neuron.
nanometer One twenty-five millionth of an inch.
nematocyst Stinging structure in some cells of corals, sea anemones, and jellyfish.
neuron A nerve cell.
neurotransmitter Chemical that carries a nerve signal across a synapse.
nodes of Ranvier Tiny gaps in the myelin sheath across which nerve signals jump.
nuclear envelope Membrane surrounding the nucleus.
nucleus Organelle that contains a eukaryote cell's DNA.
oncogene Gene that causes cancer.
organelle Membrane-lined structures inside eukaryote cells, such as the nucleus.
paracrine signaling Short-distance signaling between cells.
peroxisome Organelle that breaks down toxins.
phloem Plant tissue that carries dissolved sugars.
photosynthesis The conversion of water and carbon dioxide into sugars in plants, using the energy of sunlight.
phragmoplast Structure on which cell walls form during cell division in plants.
plasmodesmatum Structure linking two plant cells that allows water and small molecules to move from one to the other.
prokaryote cell Cell of a bacterium, which does not contain organelles.
protein Molecule formed by amino acids in the ribosome.
proto-oncogene Gene that controls the timing of each stage of the cell cycle.
red blood cell Cell that carries oxygen and carbon dioxide around the body.
ribosome Granule on which protein production occurs.
rigor mortis Stiffness of the muscles after death.
rough endoplasmic reticulum Organelle on the surface of which ribosomes occur.
sarcomere Unit of muscle that contracts.
smooth endoplasmic reticulum Organelle that packages proteins ready for export from the cell and produces some lipids.
sperm Male sex cell.
spindle Cagelike structure that forms during cell division along which the chromosomes align and move.
stem cell Cell that has yet to differentiate (change) into a particular type of cell.
synapse Gap between the ends of two neurons.
transpiration Process of water loss at the leaves of a plant.
tumor A mass of cells started by a single cell that divides uncontrollably; can be benign or malignant (cancerous).

tumor-suppressor gene Gene that inhibits cell division.

turgor pressure Pressure exerted by water-filled cavities in plant cells that keeps plants erect.

xylem Plant tissue through which water is transported.

zygote An egg fertilized by a sperm that will develop into a new organism.

American Society for Cell Biology
8120 Woodmont Avenue, Suite 750
Bethesda, MD 20814-2762
(301) 347-9300
Web site: http://www.ascb.org
The American Society for Cell Biology is an international community of biologists dedicated to advancing scientific discovery, advocating sound research policies, improving education, promoting professional development, and increasing diversity in the scientific workforce.

Canadian Bioethics Society
561 Rocky Ridge Bay NW
Calgary, AB T3G 4E7
Canada
(403) 208-8027
Web site: http://www.bioethics.ca
This forum for professionals is interested in sharing ideas relating to bioethics and in finding solutions to bioethical problems.

Human Genome Project
Oak Ridge National Laboratory (ORNL)
P.O. Box 2008 MS6335
Oak Ridge, TN 37831-6335
(865) 576-7658
Web site: http://www.ornl.gov/sci/techresources/Human_Genome/home.shtml
The Human Genome Project (HGP) was a 13-year project completed in 2003 that was coordinated by the U.S. Department of Energy and the National Institutes of Health to map the entire human genome.

Journal of Cell Biology
Rockefeller University Press
1114 First Avenue
New York, NY 10065-8325
(212) 327-8581
Web site: http://jcb.rupress.org
The *Journal of Cell Biology* is an international peer-reviewed journal owned by Rockefeller University and published by the Rockefeller University Press.

Scitable
Nature Publishing Group
25 First Street, Suite 104
Cambridge, MA 02141
Web site: http://www.nature.com/scitable/topic/cell-biology-13906536
Scitable is a free science library and personal learning tool brought to you by Nature Publishing Group, the world's leading publisher of science.

WEB SITES

Due to the changing nature of Internet links, Rosen Publishing has developed an online list of Web sites related to the subject of this book. This site is updated regularly. Please use this link to access the list:

http://www.rosenlinks.com/CORE/Cell

Alberts, Bruce, Alexander Johnson, Julian Lewis, and Martin Raff. *Molecular Biology of the Cell: 5th Edition*. Garland Science, New York, NY: 2008.

Arato, Rona. *Protists: Algae, Amoeba, Plankton, and Other Protists*. New York, NY: Crabtree Publishing Company, 2010.

Ballard, Carol. *Cells and Cell Function*. New York, NY: Rosen Central, 2010.

Collins, Francis S. *The Language of Life: DNA and the Revolution in Personalized Medicine*. New York, NY: Harper, 2010.

Cregan, Elizabeth and Bradford Kendall. *Pioneers in Cell Biology*. Minneapolis, MN: Compass Point Books, 2010.

Hodge, Russ. *Developmental Biology: From a Cell to an Organism*. New York, NY: Facts on File, 2009.

Hodge, Russ. *Human Genetics* (Genetics and Evolution). New York, NY: Facts on File, 2010

Hollar, Sherman. *A Closer Look at Biology, Microbiology, and the Cell*. New York, NY: Britannica Educational Publishing, 2012.

Kite, L. Patricia. *Unicellular Organisms*. Chicago, IL: Raintree, 2009.

Latham, Donna. *Cells, Tissues and Organs*. Milwaukee, WI: Raintree, 2008.

Maddox, Brenda. *Rosalind Franklin: The Dark Lady of DNA*. New York, NY: Harper Collins, 2002.

McManus, Lori. *Cell Function and Specialization*. Milwaukee, WI: Raintree, 2008.

McManus, Lori. *Cell Systems*. Portsmouth, NH: Heinemann Educational Books, 2010.

Panno, Joseph. *Animal Cloning*. New York, NY: Facts on File, 2010.

Panno, Joseph. *Gene Therapy*. New York, NY: Facts on File, 2010.

Schultz, Mark. *The Stuff of Life: A Graphic Guide to Genetics and DNA*. New York, NY: Hill and Wang.

Smith, Terry L. *The Evolution of Cells*. New York, NY: Chelsea House, 2012.

Somerville, Barbara Ann. *Cells and Disease*. Portsmouth, NH: Heinemann Educational Books, 2010.

Thomas, Christopher Scott. *Stem Cell Now*. New York, NY: Plume, 2006

Walker, Denise. *Cells and Life Processes*. London, England: Evans Brothers, Ltd., 2010.

N

O

P

R

S

T

U

V

W

X

Z

PHOTO CREDITS

Cover, p. 3 Cultura Science/Rafe Swan/Oxford Scientific/Getty Images; p. 6 Francis Demange/Gamma-Rapho/Getty Images; p. 7 Biophoto Associates/Photo Researchers/Getty Images; p. 8 Ed Reschke/Peter Arnold/Getty Images; pp. 12, 17 Steve Gschmeissner/Science Photo Library/Getty Images; p. 13 Feng Yu/Shutterstock.com; p. 14 M I Walker/Photo Researchers/Getty Images; p. 15 Steve Gschmeissner/Science Source; p. 18 Spike Walker/Stone/Getty Images; p. 20 Image Source/Getty Images; p. 22 Manfred Kage/Science Source; p. 33 Dr. David Phillips/Visuals Unlimited/Getty images; p. 35 BSIP/Universal Images Group/Getty Images; p. 40 Ed Reschke/Peter Arnold/Getty Images; p. 42 (top) Biophoto Associates/Photo Researchers/Getty Images; p. 42 (bottom) Visuals Unlimited, Inc./Dr. Peter Siver/Getty Images; p. 43 Kallista Images/Getty Images; p. 45 Diego Cervo/Shutterstock.com; pp. 46, 50 iStockphoto/Thinkstock; p. 52 Marcus Brandt/AFP/Getty Images; p. 54 lakeemotion/Shutterstock.com; p. 55 D.P. Wilson/FLPA/Science Source; p. 56 Ulrich Baumgarten/Getty Images; p. 58 Blend Images/ERproductions Ltd/the Agency Collection/Getty Images; p. 63 Chris McGrath/Getty Images; p. 64 National Cancer Institute; p. 67 Vanderlei Almeida/AFP/Getty Images; p. 71 Dr. Cecil Fox/National Cancer Institute; p. 73 BSIP/Universal Images Group/Getty Images; p. 74 Science Source; p. 75 Last Refuge/Robert Harding World Imagery/Getty Images; p. 77 Science Source/Photo Researchers/Getty Images; p. 78 Universal Images Group/Getty Images; p. 79 Central Press/Hulton Archive/Getty Images; p. 80 (right) SMC Images/The Image Bank/Getty Images; p. 81 James King-Holmes/Science Source; interior pages background images © iStockphoto.com/Julián Rovagnati (borders), © iStockphoto.com/Sergey Yeremin; top of pp. 6, 12, 22, 34, 46, 54, 64, 74 © iStockphoto.com/aleksandar velasevic; all other photos and illustrations © Brown Bear Books Ltd.

JAN 2014
Received
Salinas Public
Library